AF553939

POLARIZATION AND INTERFEROMETRIC

POLARIZATION AND INTERFEROMETRIC

By
Dr. K. Sonamuthu
Reader
Deptt. of Physics
J.N.R. Mahavidyalaya
Port Blair
Andaman and Nicobar Islands

DISCOVERY PUBLISHING HOUSE PVT. LTD.
NEW DELHI-110 002

First Published - 2010

Reprinted - 2017

ISBN: 978-81-8356-593-6

Polarization and Interferometric

Published by:

DISCOVERY PUBLISHING HOUSE PVT. LTD.

4383/4B, Ansari Road, Darya Ganj

New Delhi-110 002 (India)

Phone: +91-11-23279245, 43596064-65

Fax: +91-11-23253475

E-mail: discoverypublishinghouse@gmail.com

sales@discoverypublishinggroup.com

web: www.discoverypublishinggroup.com

Printed at:

Infinity Imaging Systems

Delhi

Preface

The book titled "Polarization and Interferometric", here I have made an attempt to know and rectify that the phenomenons that takes place against to the light, for the light and of the light. That is known as dispersion. Several dispersion effects or encountered by the pulse of light propagation through a fibre and these acts to spread out of the pulses in the time domain by changing the shape of the pulse. As a result pulses may be got merged in the previous or in the succeeding pulses became indistinguishable at the receiver end. Hence here I have tried to reduce the PMD by a simple technique called Interferometry. It is the auto correlation of the optical electrical field with respect to time using the interference pattern created by adding a wave to a delayed wave. This book deals with that how the Interferometer splits an incoming light wave along its path and recombined the light at the output.

Finally it is concluded that by this method an increasing the wavelength of a coherence source the PMD is reduced by means of which the speed of the communication of optical signal can be increased through optical fibre.

Author

Contents

1

Optical Fibers in Optical Communication Systems

INTRODUCTION

Optical communication systems date back two centuries, to the "optical telegraph" that French engineer Claude Chappe invented in the 1790s. His system was a series of semaphores mounted on towers, where human operators relayed messages from one tower to the next. It beat hand-carried messages shads down, but by the mid-19th century it was replaced by the electric telegraph, leaving a scattering of "Telegraph Hills" as its most visible legacy.

Alexander Graham Bell patented an optical telephone system, which he called the Photophone, in 1880, but his earlier invention the telephone, proved for more practical. He dreamed of sending signals through the air, but the atmosphere didn't transmit light as reliably as wires carried electricity. In the decades that followed, light was used for a few special applications.

In the intervening years, a new technology slowly took root that would ultimately solve the problem of optical transmission, although it was a long time before it was adapted for communications. It depended on the phenomenon of total internal reflection, which can confine light in a material surrounded by other materials with lower refractive index, such as glass in air. In the 1840s, Swiss physicist Daniel Collodon and French physicist Jacques Babinet showed that light could be guided along jets of water for fountain displays. British physicist John Tyndall popularized light guiding in a demonstration he first used in 1854, guiding light in a jet of water flowing from a tank. By the turn of the century, inventors realized that bent quartz rods could carry light, and patented them as dental illuminators. By the 1940s, many doctors used illuminated plexiglass tongue depressors.

Optical fibers went a step further. They are essentially transparent rods of glass or plastic stretched so they are long and flexible. During the 1920s, John Logie Baird in England and Clarence W.Hansell in the United States patented the idea of using arrays of hollow pipes or transparent rods to transmit images for television or facsimile systems. However, the first person known to have demonstrated image transmission though a bundle of optical fibers was Heinrich Lamm, than a medical student in Munich. His goal was to look inside inaccessible parts of the body, and in a 1930 paper the reported transmitting the image of a light bulb filament through a short bundle. However, the unclad fibers transmitted images poorly, and the rise of the Nazis forced Lamm, a Jew, to move to America and abandon his dreams of becoming professor of medicine.

Neither Van Heel nor Hopkins and Kapany made bundles that could carry light far, their reports the fiber optics revolution. The crucial innovation was made by van Heel, stimulated by a conversation with the American optical physicist Brain O'Brien. All earlier fibers were "bare", with total internal reflection at a glass-air interface. This protected the total reflection surface from contamination, and greatly reduced cross talk between fibers. The next key step was development of glass-clad fibers, by Lawrence Curtiss, then an undergraduate at the University of Machigan working part-time on a project to develop and endoscope to examine the inside of the stomach with physician Basil Hirschowitz, physicist C.Wilbur Peters. By 1960, glass-clad fibers had attenuation of about one decibel per metre, fine for medical imaging, but much too high for communications.

Meanwhile, telecommunications engineers were seeking more transmission bandwidth. Radio and microwave frequencies were in heavy use, so they looked to higher frequencies to carry loads they expected to continue increasing with the growth of television and telephone traffic. Telephone companies though video telephones lurked just around the corner, and would escalate bandwidth demands even further. The cutting edge of communications research were millimetre-wave systems, in which hallow pipes served as waveguides to circumvent poor atmospheric transmission at tens of gigahertz, where wavelengths were in the millimetre range.

Even higher optical frequencies seemed a logical next step in 1958 at Alec Reeves, the forward looking engineer at Britian's Standard Telecommunications Laboratories who invented digital pulse-code modulation before World War-II. Other people climbed on the optical communications bandwagon when the laser was invented in 1960. Optical fibers had attracted some attention because they were analogous in theory to plastic dielectric waveguides used in certain microwave applications. In 1961, Elias Snitzer at American Optical, working with Hicks at Mosaic Fabrications, demonstrated the similarity by drawing fibers with cores. So they carried light in only one waveguide mode. However virtually everyone considered fibers too lossy for communications; attenuation of a decibel per metre was fine for looking September 1970, they announced they had made single-mode fibers with attenuation at the 633 nanometer helium-neon below 20 dB/Km. The fibers were fragile, but tests at the new British Post Office Research Laboratories facility in Martlesham Heath confirmed the low loss.

The Corning breakthrough was among the most dramatic of many developments that opened the door to fiber-optic communications. In the same year, Bell Labs and a team at the Ioffe Physical Institute in Leningrad made the first semiconductor diode lasers able to emit continuous wave at room temperature. Over the next several years, fiber losses dropped dramatically, aided both by improved fabrication methods and by the shift to longer wavelengths where fibers have inherently lower attenuation [1, 2].

GENESIS OF FIBER OPTICS

A light source a light guide and a light detector constitute the principal elements of a fiber optic transmission system. The light source receives a coded electrical signal (such a voice, data or video information) transforms it into a light signal, and steers this signal on to the light guide. The light signal is confined along the light guide for transmission to the detector, where it is received and reconverted into the original electrical message. Although this concept is simple, it emerged only within the past 15 years.

The visible portion of the electromagnetic spectrum, namely light, has been used for centuries, as in the case of hilltop beacons, smoke signals (reflected light) and even Alexander Graham Bell's 1980 photo phone, which was nothing but a device that modulated the incoming sunlight and reflected it over several hundred metres for detection and demodulation back into speech.

The limitations of such line-of-sight communication systems lies in the fact that the light signal is unguided; instead it would be transmitted through the atmosphere, encountering distortion problems, even when weather positions are fair. Such signals can be scatted or blocked entirely by physical obstacles; also, in the case of messages which are confidential, the signal can be intercepted and even deciphered with a knowledge of the code used. So a viable light-wave communication system needs a light guide so as to preserve the signal and hence increase the reliability and distance of signal transmission.

Development of a Light Guide

Even though the possibility of guiding light had been demonstrated as early as in 1854, in the form of reflections in a curved stream of water coming out of a hole in the side of a pail, the concept of a practical light and viable guide emerged only in 1910 in the form of a solid cylinder capable of guiding a wide range of electromagnetic waves, including those of visible light. Another possible light guide emerged in the form of a hollow tube with a highly reflective metal coating on its inner surface. In either case light injected into one end of the guide, bounded back and forth; the high signal loss,· of these devices, occurring when light had to pass through curved paths to change direction, rendered such devices impractical. Another device was in the form of a heated gas-filled conduit; but it was also not suitable for commercial applications on account of its disadvantages of complexity and high cost.

In the 1930s experiments were conducted on a guide comprising simple filaments of glass fiber packed into bundles. In a time span of 20 years, these glass fiber bundles were being used as light conduits for card readers; later they were used in medical endoscopes. At this time, optical fibers also began to appear in photography where, fused together, the glass fiber bundles were used as face plates in image intensifiers and field flatteners. Each fiber had become a composite structure. The core region, with its high refractive index, carried most of the light and was surrounded by a cladding that had a lower refractive index. Although this design improved the transmission efficiency as well as the handling capacity of each fiber, and also reduced the cross talk amongst the fibers in a bundle, signal losses restricted the optical fiber to applications of utmost a few metres.

By 1970, leading American glass manufactures were able to produce fiber, of sufficient purity, for use in the telecommunication industry with an attenuation of less than 20 dB/km. The fiber was fabricated by using a method of synthesizing silica glass. The raw materials were vaporized and deposited inside a length of quartz glass tubing, which was then solidified into a rod and drawn in the form of a fiber.

TRANSMITTER AND RECEIVER ELEMENTS

Apart from optical fiber, an optical transmission system would need, in a practical system, two other key elements, namely, the transmitter and receiver which had been already developed independently.

In order to send a light signal through optical fiber, it is required to have a source that emits a bright directional signal. Also, since light signals are often in digital form (that is, the information is conveyed in the form of series of light pulses), the light sources should be capable of being turned on and off quickly. The more rapid the modulation, the more would be the information conveyed in a given time.

By 1970, the different technologies, underlying the complete fiber optic transmission system, had coalesced. The following decade saw research institutions as well as manufactures of telecommunications equipment seriously engaged in efforts to refine the components and assemble them into practical optical transmission systems.

With the potentiality of the glass fiber was unlimited information carrying capacities at certain wavelengths, the Bell Laboratories developed a laser with a promise of continuous operation at room temperature for more than a thousand hours. Other phenomenal developments followed, including the fabrication of a commercial optical cable, a reliable high power LED as well as prototype splice and connector, a fiber optic avalanche photo-diode detector and a fiber link installed on a destroyed. By 1978, the Japanese had succeeded in reducing fiber losses to a low figure of 0.2 dB/km. At the same time several fiber systems were being processed and undergoing field testing, so as to be used in telephone systems. Currently, hundreds of fiber optic transmission systems, of experimental and operational types, are in use around the world.

Fiber Optics Transmission Systems

While radio and copper wire transmission systems have been in vogue for along time, light wave communication has sparked renewed interest all over the world by different telecommunication organizations. This time the transmission medium is not air (which is a poor light conductor) but cables optical fibers. In the last twelve years, the fiber optics technology has advanced to the point where practical fiber systems present viable alternatives to copper in telecommunications.

Basically, fiber optic transmission arose from a synthesis of two hitherto unrelated technologies: the semiconductor technology and the optical waveguide technology. The former technology would supply the raw materials for the light sources and detectors needed in an optical transmission system, while the second technology would give rise to a transmission medium optical fiber. In the late 1960's it was realized that a combination of the two technologies would lead to a transmission system that would have certain inherent advantages over conventional copper systems, especially in the case of the digital systems of the developing countries.

Apart from having lower transmission losses than copper, the fiber provides higher bandwidth. These two basic advantages would give systems with longer repeater spacing and higher information carrying capacities than are possible with copper. The smaller size and lower weight of fiber help to make cables that are easier to be handled and installed; in addition, the fibers themselves enjoy immunity to electromagnetic interference such as lightening, ac induction, and cross talk from other fibers in a cable. The most attractive feature of the fiber lies in the availability of abundant and inexpensive silica material used to make the fiber for use in attractive transmission systems of low cost.

Fiber in Telecommunication Systems

Communication, between switching office, is normally via the trunk facilities. Trunk transmission in the network has been primarily of the analogue type until 1962 when the DSI rate (1.544 Mb/s) of digital transmission was introduced. Since then the DSI rate copper systems have multiplied and are now the standard carrier systems, especially in 10 km - 320 km trucking applications. The successof these systems was due to their better performance and less cost as compared to the systems they replaced. In this context, fiber optic systems would be very competitive and viable among the next generation trunk transmission systems. As shown in Fig.1.1.

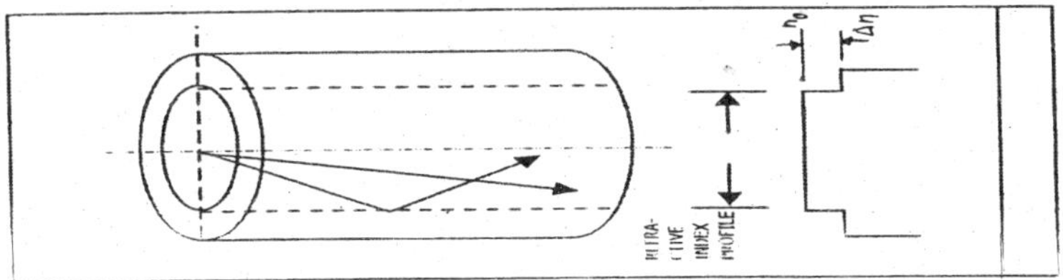

Fig. 1.1: Fiber in Communication

OPTICAL FIBER IN SHORT DISTANCE COMMUNICATION

Numerous techniques [3, 4, 5] are developed for fabrication of glass fibers so as to get better optical properties suitable for optical communication, and also to combat with ever increasing demand of fiber in diverse application areas. Various processes are being modified such as perform making, melting process, melting environment, homogenization/fining, and reactive atmosphere processing [6]. Hermetic carbon coating has been shown to be effective barrier for water and hydrogen [7, 8, 9]. Polymers such as ultra violet radiation curable acrylate and Teflon-FEP are used as coating for fluoride glass optical fibers [10]. However, these materials have high permeability of water and hence are not adequate for fabrication of high strength optical fibers. Oxide glass has low permeability of water than fluoride glass. Efforts are also made to fabricate plastic optical fibers as a compliment for glass fibers in short distance communication links [11]. The fabrication and characterization of side hole single mode optical fibers have been situated [12]. The thermal modeling is reported for optical fibers drawing process to see that effect of fiber diameter and drawing speed on temperature distribution [13]. The most fundamental physical parameter for optical fiber is geometry, since the dimension of the fiber determines its ability to splice and to be terminated. Therefore, when the fiber is fabricated, it is important to study the above mentioned parameters, which are useful in assessing the communication performance of the fiber.

Fiber Fabrication

At first it seemed that this low-loss transmission requirement would be very difficult to achieve; but finally tremendous advances have been made in several laboratories. The Southampton University held the world record for the best fiber, with a loss of only 5.8 dB/Km, in a configuration consisting of a fine glass capillary tube as the cladding filled with a special liquid (hexachlorobutadiene) as the core. Other laboratories in the USA improved on this result using solid fibers consisting of silica doped with titania, germania or boric oxide as one component and pure silica as the other. Again, the Southampton University produced another new type of fiber using a rather unexpected material namely phosphosilicate glass.

Preform Fabrication

Optically transparent oxide glasses fall into two main types: multi-component glasses and high silica content glasses. The fabrication processes, for these glasses, are different, since high silica glasses require a higher fabrication temperature than multi-component glasses; glass objects are made of multi-component type. The attenuation and dispersion, of multi-component glass and of plastic fibers, are too high to be practical for most telecommunication applications.

Fiber Drawing

The performance of optical fibers, specifically the transmission losses, bandwidth and strength, is strongly influenced by the fiber drawing and coating process. This process must be treated as an integrated operation designed to impart desired properties to the resulting fiber which consists of feed glass, heat source, fiber diameter monitor, cooling system, coating applicator, curing apparatus, and fiber puller/winding mechanism. The components are arranged in vertical position to take advantage of gravitational forces. The feed glass perform is fed at speed of 0.02-0.03 cm/s depending on the heat source, perform diameter and draw speed. Continuous lengths of more than 40 km have been drawn from such a perform in one drawing. Fiber drawing proceeds by heating the tip of the perform to a molten state and allowing it to extend under the force of gravity downward into a fine-diameter filament whereas pulling and winding mechanism sustains force. The fiber diameter is monitored continuously as the fiber passes through the furnace. Any variation of the fiber diameter is detected by a control device that automatically readjusts the temperature and the speed of pulling mechanism to compensate for initial change.

Since this bare fiber consists of core and clad, the fiber diameter is an important parameter, which influences wave propagation phenomenon used in optical fiber. When the fiber comes out of the furnace, it is cooled in the presence of helium gas without altering its physical dimension. Then it enters the coating applicator immediately.

FIBER OPTICS TECHNOLOGY

Light Through Fibers

Let us consider a construction Fig.1.2 with n_1 and n_2 being refractive indices of two materials, i.e. the core and cladding surrounded by air of refractive index *n*.

Assuming that $n_1 > n_2$. In the case of ray1, the incident ray strikes the core and cladding interface at an angle $\theta_i > \theta_c$ (critical angle) and is totally internally reflected in the core. Ray 2 strikes the interface at an angle $\theta_i < \theta_c$ and is refracted into the cladding. But it again strikes the cladding-to-air surface $\theta_i > \theta_c$ here θ_c is the critical angle for cladding-to-air surface. This ray will reflect back into the cladding, then to the core and so on.

In the case of ray 3, $\theta_i < \theta_c$ and $\theta_i < \theta_c$; the ray is from the core to the cladding and to the air.

Ray 1 is the most important ray from fiber-optics points of view. However ray 2 also propagates through the length of such a construction; but scattering and absorption are very high in the irregular cladding material so as to result in high loss as compared to the one through the core. Since some of the energy will always travel through the material outside the cores, cladding is provided instead of air surrounding the core. However, the main effort of the designer will be to trap the energy in the core to the maximum extent.

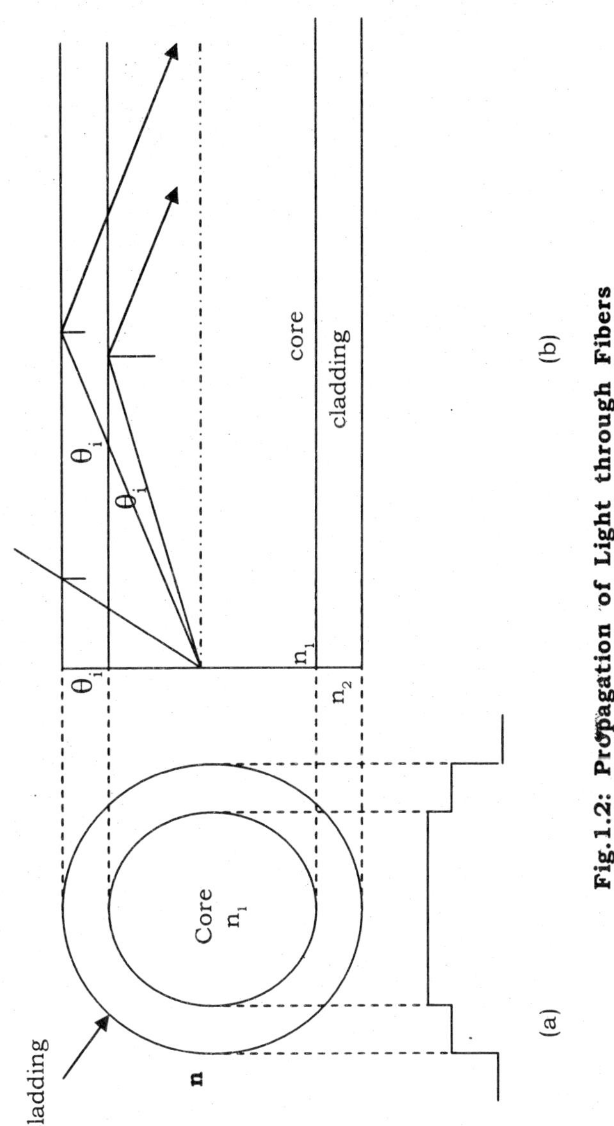

Fig.1.2: Propagation of Light through Fibers

This type of construction of the fiber is known as step-index fiber; in the case refractive indices, of the materials, would vary in discrete steps n_1 and n_2.

Numerical Aperture

Let us consider a ray entering the fiber and making an angle θ with the axis. After refraction, it emerges at an angle ϕ. Also, at the core-cladding interface angle, we have $\theta_i < \theta_c$

It follows, from Smell's law, that

$n_1 \sin\theta = n_1 \sin\phi$

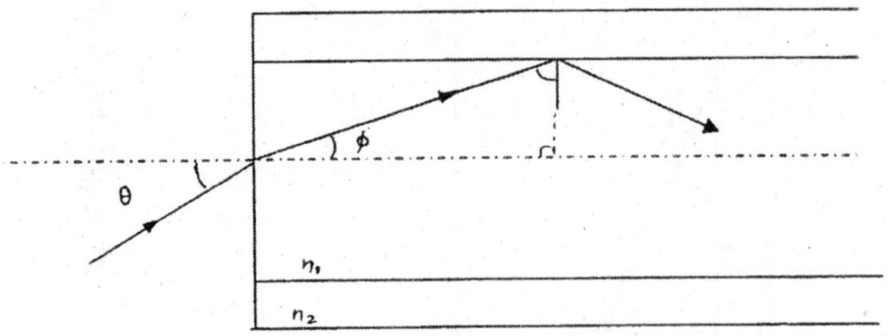

Fig.1.3: Refraction of Light through Fiber

But $n_1 \sin\phi = n_1 \sin(90 - \theta_i) = n_1 \cos\theta_i$

So that $n \sin\theta = n_1 \cos\theta_c$

When $\theta_i = \theta_c$

$n \sin\theta = n_1 \cos\theta_c,$

But we know that $\sin\theta_c = \dfrac{n_2}{n_1}$

So that $\cos\theta_c = \dfrac{n_1^2 - n_2^2}{n_1}$

Hence

$$n \sin\theta = n_1, \frac{n_1^2 - n_2^2}{n_1}$$

= Numerical Aperture of the fiber.

With n=1 (for air)

We have N.A= $\sin\theta = (n_1^2 - n_2^2) = \theta$ (for small angles).

In order to have $\theta_I > \theta_c$ (for total internal reflection), ϕ or (in turn) θ must be small.

The quantity, N.A. = $\sin\theta = (n_1^2 - n_2^2)$, defines the maximum acceptable light cone for a fiber of the given refractive index; this quantity is a very important parameter of the fiber.

Multimode and Monomode Fibers

When the core diameter, of the fiber [Fig. 1.4], is large, multimode propagation is possible; under such conditions; a fiber is referred to as to multimode fiber. The disadvantage of such a fiber would be its wide dispersion. With the dependence of dispersion on *a*, it is possible to reduce the dispersion of the fiber, if its diameter is reduced to an extent that a single mode can be propagated; in that case, this type of fiber would be known as a *mono-mode fiber.*

Grade Index Fiber

Another solution, to the problem of mode or multi-path dispersion is obtained by a graded-index fiber whose refractive index is not constant along the radius of the core but varies according to a certain present relation from the core centre towards the outer diameter, such as the core index decreasing along the radius from the center. Since the core center has a higher refractive index, the velocity of the light wave

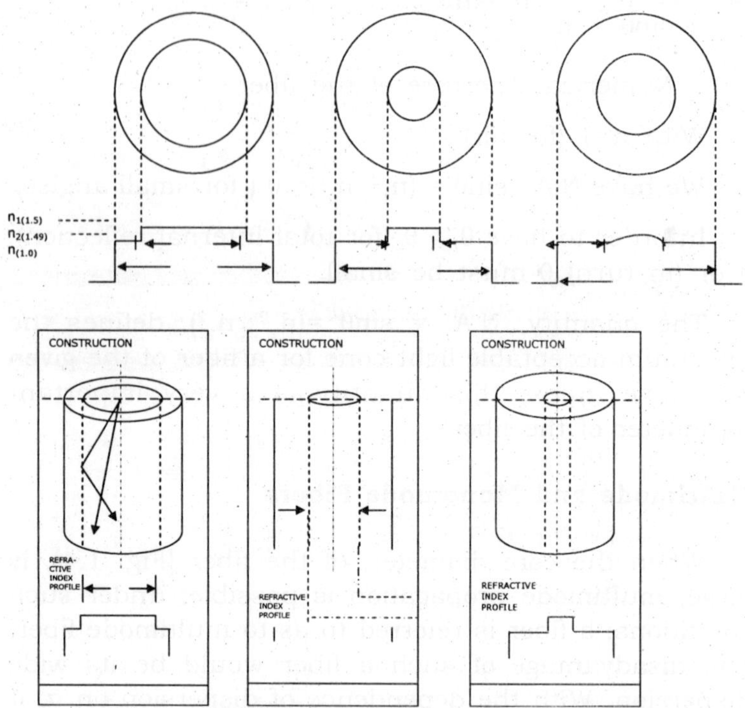

Fig. 1.4 Multimode Propagation in Fibers

(V=c/n) near the center is lower than that near the surface; also, due to a constant reduction in the value of refractive index along the cross-section, the ray will take a curved path as show in Fig. 1.4. The above two factors give a very much reduced amount of mode dispersion.

It can be shown that, if the refractive index follows a parabolic profile,

Then $n(x)=n_1[1-\Delta(x/r)]^2$, $0 < x < r$ (for the core)

and $n(x)=n_2=n_1[1-\Delta]$, (for the cladding)

Here r = radius, x = distance measured along the radius from the centre and

$$\Delta = \left(\frac{n_1 - n_2}{n_1}\right)$$

All off-axis rays would follow a sinusoidal path. By working out a solution with such sinusoidal paths, it can be shown that a parabolic refractive index would lead to zero mode dispersion ideally.

In general

$n(x) = n_1[1- \Delta(x/r)]\alpha$, for $0 < x < r$

and

$n(x) = n_2 = n_1(1- \Delta)]$ for $x > r$

if $\alpha = \infty$

then $n(x) = n_1$, for $0 < x < r$

and $n(x) = n_2 = n(1- \Delta)]$ for $x > r$

This applies to a step-index fiber.

This optimum value of α in practice depends upon the materials used for the fiber; this α factor is very important to be controlled during the manufacture of the fibcr.

The very small difference, between n_1 and n_2 implies that the axis angle, for a core-trapped ray, is small. Such fibers are referred to as weakly guided structures. The multimode step index fiber is useful for only short transmission lengths because of its large dispersion, but provides a higher coupling efficiency due to its large core cross-section. The mono-mode fiber, on the other hand, overcomes the problem of multi-path dispersion, but has the disadvantage of a very small cross-sectional area. The graded-index fiber, however, exhibits relatively loss-mode dispersion, with a relatively large core area.

ADVANTAGES OF OPTICAL FIBERS

In addition to the advantages of having extra information bandwidth using light as carrier signal, the optical fiber communication systems have several other advantages over the conventional systems.

They are as follows:-

(i) By virtue of its manufacturing material, it is very light and easy to handle.

(ii) Because of very low transmission loss, the distance between repeater stations increased greatly

(iii) Optical fibers can be used in explosive as well as high voltage environment due to the absence of any hazards arising out of short circuits.

(iv) It is ideal for secret communication systems because it is very difficult to tap but very easy to monitor.

(v) They are virtually unaffected by outdoor atmospheric conditions.

(vi) Since the only carrier in the fiber is light, there is no possibility of a spark from a broken fiber.

(vii) It will not corrode and is unaffected by mot chemicals. It can be buried directly in most kinds of solids or exposed to most corrosive atmospheres in chemical plants.

(viii) They are immune to electromagnetic interference and avoid cross talk.

FUTURE FOR FIBER OPTICS

With the advent of the first generation fiber transmission systems in the market, fiber research and development would continue. The progress, achieved in fiber research and development have been achieved by people, with highly specialized skills, who have used innovative but costly equipment to produce a limited variety of experimental fiber samples. Manufacturing, however, requires a large volume production at minimal cost. Much room for innovative engineering development lies in the translation of technology into an industrial product.

Continuous research is being directed towards bringing the fiber attenuation level over a broad range of wavelengths close to the theoretical limit, which, at the 1550-nm wavelength, is about 0.2 dB/Km. The commercial production of these low loss fibers should be possible in the near future. The loss level of these fibers can improve only slightly, since hydroxyl water has already been almost totally eliminated. Heavier-atom-halide materials, however, have the potential to raise the molecular absorption of the fiber to the 3000-5000 nm range where Rayleigh scattering and ultraviolet absorption are extremely low. The potential impact of such fibers would be properly estimated after investigating the possibility of manufacture of such fibers and associated compatible light sources, detectors, and splicing techniques. The quest, for higher bandwidths, is straighter forward.

With fibers subjected to very high levels of optical power density, there would be non-linear refractive index effects which may lead to the generation of new lasing wavelengths, wavelengths filtering and pulse compression. Investigations have already been launched to study the importance of such effect as well as fiber sensors for the detection of thermal, acoustic, mechanical and electromagnetic signals. These properties may contribute to future forms of fiber optics for tele-communications.

SUGGESTIONS FOR FURTHER WORK

Using the same order in which topics were discussed in this thesis, the following future work suggestions are made. From the modeling point of view, an analytical derivation of the correlation bandwidth would give confidence to the empirical result. Alone with that, analytical derivations of the simulation results pertaining to variance on the mean and cross-correlation error would be desirable.

The measurements conducted in this thesis are laborious. For a long term PMD characterization test-bed, the processes of (a) determining the correlation temperature of a device and (b) generating independent PMD instances for a device and measuring the mean DGD should be automated, now that a methodology is in place. This would help narrow down the uncertainty on a device's mean DGD. In terms of statistical model characterization, automated measurements would permit a population of DGD samples large enough to characterize the tail of the distribution to be more easily obtained over a given period of time. the would help ascertain to a further degree the theoretical Maxwellian distribution.

It would appear that the possibilities of numerical simulations are endless. Form this perspective, a pulse of given duration could be made to propagate through the fiber already created by the simulation program. After addition of this feature to the program, the pulse distorting effect could be examined, and consequential communications penalty derived and compared to that of the literature. A study of PMD compensation could ensue from this.

Finally, given the limitations of the devices which were measured, and the knowledge that the major DGD contributors were devices which are usually passive, a test of a system with high PMD fiber spans would be ideal. In this case, the DCMs could be held at constant temperature, and only the PMD in fiber spools would have to vary over time (or temperature). Due to the high DGD, fewer measurements would have to be taken to obtain mean DGD estimates with the same confidence interval as obtained in this thesis. From the systems perspective, a link with high DGD could be used to further study the effects of first and second-order PMD over time.

APPLICATIONS

The Effects of Polarization-mode Dispersion on the Phase of the Recovered Clock

The impact of polarization-mode dispersion (PMD) on the phase of the recovered clock in the receiver is analyzed. The effects of first-order PMD on different clock-recovery configurations utilized for return-to-zero (RZ) and non-RZ (NRZ) formats are studied. Closed-form expressions relating the PMD-induced sampling time shift with the differential group delay and the power ratio between the principal states of polarization are obtained for each high-<formula formulatype=”inline”><tex>Q </tex></formula>filter-based clock recovery. An experimental validation at 10Gb/s is also shown for the case of NRZ data format.

Signal Propagation over Polarization - Maintaining Fibers: Problem and Solutions

Polarization-maintaining (PM) fibers are able to preserve the state of polarization (SOP) of a signal in the fiber reference frame. The SOP follows one of the axes of the fiber defined by the mechanical constraint that has been inserted during manufacturing. This ability is characterized by the polarization extinction ratio, which should be as high as possible. It is known to be very high (often above 50 dB) for the fiber itself. However, the somewhat law polarization extinction ratios (often between 20 and 30 dB) of connections (connectors, splices, <formula formulatype = “inline”> <tex>$1dots$</tex> </formula>) induce strong distortions of the signal. The authors first explain why for one PM patch cord, i.e., a fiber and its connectors, with numerical and analytical models. They then extend these domesl to a patch-cord concatenation. Finally, they show and compare analytically and numerically different solutions to this problem: fiber length reduction, polarizer insertion, high-performing connectors, and axis alternation. The latest solution consisting in alternating the slow and the fast axes from one fiber to the next one is particularly efficient. They have implemented these solutions for the purpose of the source of the & Idquo; Laser M & eacute; gaJoule, & rdquo; and it exhibits far better performance.

Performance Fluctuations in Submarine WDM Systems

Polarizations-related effects are major source of performance fluctuations in long-haul systems. Understanding the system implications of polarization-dependent loss, polarization mode dispersion, linear and nonlinear channel crosstalk, polarization-dependent gain, and polarization hole burning is important for a successful system design and operation. In this paper, the polarization and nonlinearity-driven performance fluctuations in 10-Gb/s wavelength-division multiplexing submarine transmission systems, including repeated and unrepeated systems, are analyzed. The proposed accumulated over a long period on deployed repeatered systems. Furthermore, the case of 40-Gb/s transmission is briefly discussed.

Location of High PMD Sections of Installed System Fiber

This paper describes two techniques for the locations of high polarization-mode dispersion (PMD) sections of installed system fiber. The first technique uses the recovery of local linear birefringence by polarization backscatter reflectometry, whereas the second uses the wavelength dependence of the backscattered signal to determine the spatial distribution of the PMD.

REFERENCES

1. Fleming JW , "Fiber Optics Technicians Hand Book" by JIM Hayes, Delmar Publ. New York, Electron Lett. 14, (326), 1978.
2. Jeff Hecht, "The Story of Fiber Optics", Oxford University Press, New York, 1999.
3. Beals K J., Day CR, Duncan WJ, Dunn AG, Dunn PL, Newns GR and Wright JV, 5th European Conference on optical Fiber Communication (New York: McGraw Hill), 1979.
4. Izawa T., Sudo S., and Hanawa F ., Trans. IEEE 62, (779) 1979.
5. MacChesney JB Proc. IEEE 68(1181), 1980.
6. France Paul, Caster SF Moore W. and Williams JR in "Optical Fiber Material and Processing", MRS Symp. Proc. 172 (134), 1990.
7. Lomair PJ, Kranz KS, Walker KL and Huff RG *Elect. Lett.* 24(132),1988.
8. Huff RG, Dimarcello FV and Hast Jr AC, *Tech. Digest* OFC Paper TUG 2, 1989.
9. Lu K. E., Lee MT Powers D. R. and Glassman S., in *Tech. Digest,* PDI,OFC (425), 1989.

10. Ihishi V., Fujiura K. and Takahashi S., MRS Symp. Proc, 1990.

11. Izawa T., Shabita N. and Takeda A., *Appl. Phys. Lett.* 3133, 1977.

12. Izawa T. and Sudo S., *Optical Fiber Materials and Fabrication* (36), 1987.

13. Papamichel H. and Miaoulis *J., MRS Symp. Proc.* 172(43), 1990.

14. *Journal of Lightwave Technology*, 2006.

2

Effect of Dispersion on Optical Fibres

INTRODUCTION

It may be noted that a light pulse with a given width and amplitude injected into one end of a fiber should arrive at its other end with the shape and width unchanged, and only its amplitude is reduced by losses. If the losses are extremely large, the pulse amplitude at the receiving end will be small to be detected and a repeater has to be included to boost up the signal before its entering into next section.

Several dispersion effects are encountered by the pulse of light propagating through a fiber and these act to spread out the pulse in the time domain changing the shape of the pulse. As a result, the pulse may be got merged in the previous or in the succeeding pulses becoming indistinguishable at the receiver end. The effect is known as inter symbol interference (ISI). The pulses may be separated by increasing the time interval between pulses, but thereby the maximum bit rate will be reduced.[1]

OPTICAL FIBER AND CABLE

Analytical research, on optical fiber, began in the early 1970's. In 1974, experimental silica fibers were fabricated using a chemical vapour deposition process.

Optical Properties

An advantage of the optical fiber as a telecommunications medium is the small size of the fiber itself. With a typical diameter of only 125 microns, it possess a large variety of properties. The fiber design utilizes the fact that electromagnetic waves, in a dielectric, tend to travel in a region of high refractive index called the 'core'. In a typical fiber, this core has a graded refractive index is 50 microns in diameter and is surrounded by a cladding a glass of lower index. This cladding, in turn, is protected by a plastic jacket with an outside diameter of 330 microns.

In the currently designed fibers, light rays, that enter the core within a core having a half-angle q - 10^0 [corresponding to numerical aperture (NA) = sin q = 0.18], will propagate along the fiber. The larger the numerical aperture the more efficient would be the coupling of light from a source and the better would be the signal transmission around bends. But a large value of NA would be associated with higher attenuation and larger signal distortion [1].

The initial break-through, that permitted optical fibers to be used for telecommunications, appeared when the signal loss due to the fiber was reduced to less than 20dB per Km. This reduction resulted from the technique of holding down metallic and hydroxyl ion impurities in the fiber to a few parts per billion levels and from the ability to manufacture fibers almost totally free of structural imperfections. In the past decade, fiber-induced signal loss has been substantially reduced. So, the most recent generation of fibers can carry telecommunications signals in the wavelength range between 700 nm and 1800 nm. At the wavelengths of 1300 nm and 1550 nm, loss levels, of less than 1 dB/Km, have been achieved. Attenuation, at shorter wavelengths, is dominated by Rayleigh scattering, while attenuation, at longer wavelengths, is due to infrared absorption, the peak around 1400 nm being from hydroxyl absorption. Light transmitters and receivers are currently available for the 840 nm region. Intensive research and development, on transmitters and receivers, would help to exploit the longer wavelength regions, in the fiber, with a loss of less than 1dB/km.

As the amount of attenuation has been reduced over the years, there remains the problem of a second optical property, namely, dispersion. Dispersion causes signal pulse broadening which, in turn, limits the pulse rate or signaling frequency that a fiber can transmit. Two types of dispersion exist: chromatic dispersion and multi-mode dispersion. At a wavelength of about 840 nm, chromatic dispersion causes pulse broadening of about 0.1 ns per kilometre on fiber per nanometer of source width [2].

FABRICATION AND FIBRE DRAWING

The perform is fabricated by means of modified chemical vapour deposition (MCVD) process. Initially the substrate tube is cleaned by hydrofluoric acid and fire-polished to remove stress in the glass and later on etched with freon to remove surface impurities. The reactants ($SiCl_4$ + O_2) are introduced at one end of the rotating tube while an exhaust is located at the other end. Cladding layer is made by deposition of materials such as $SiCl_4$, G_eCl_4 and $POCL_3$ in different environments such as oxygen, helium and Freon. During clad process, Freon reduced the refractive index of the clad and reaction converts halides into oxides. Approximately, 25-30 layers of cladding are deposited on the inside of the substrate tube. A core layer is deposited by means of compounds like $SiCl_4$, G_eCl_4 in gaseous environment of oxygen, helium and Freon. $GeCl_4$, converts to GeO_2, which serves as a dopant to increase the refractive index of the core layer, whereas helium lowers the deposition temperature [4]. The flow of reactants and the speed of traversing oxyhydrogen burner are closely monitored using a video camera. When the desired thickness is achieved, valve is closed to stop the flow of the reactants. After deposition of core layers on the inner side of tube this composite tube is heated to high temperature from outside so that it collapses to form a solid rod. The freon gas is passed while collapsing so that the core has higher refractive index than the clad as well as the solid glass rod becomes free of air bubbles. The chlorine gas is also used in combination with Freon so as to remove OH from SiO_2. This perform is subjected to fiber drawing process [5] as shown in Fig. 2.1

The reaction takes place during perform fabrication.

$SiCl_4 + SO_2 \rightarrow SiO_2 + 2Cl_2$: Conversion to quartz

$GeCl_4 + O_2 \rightarrow GeO_2 + 2Cl_2$: Alteration refraction

$4POCl_4 + 3O_2 \rightarrow 2P_2O_5 + 6CL_2$: Alteration of refractive index

Freon CCl_4F_2 on hearing forms the compound SiF_4 instead of SiO_2, which lowers the deposition rate of SiO_2.

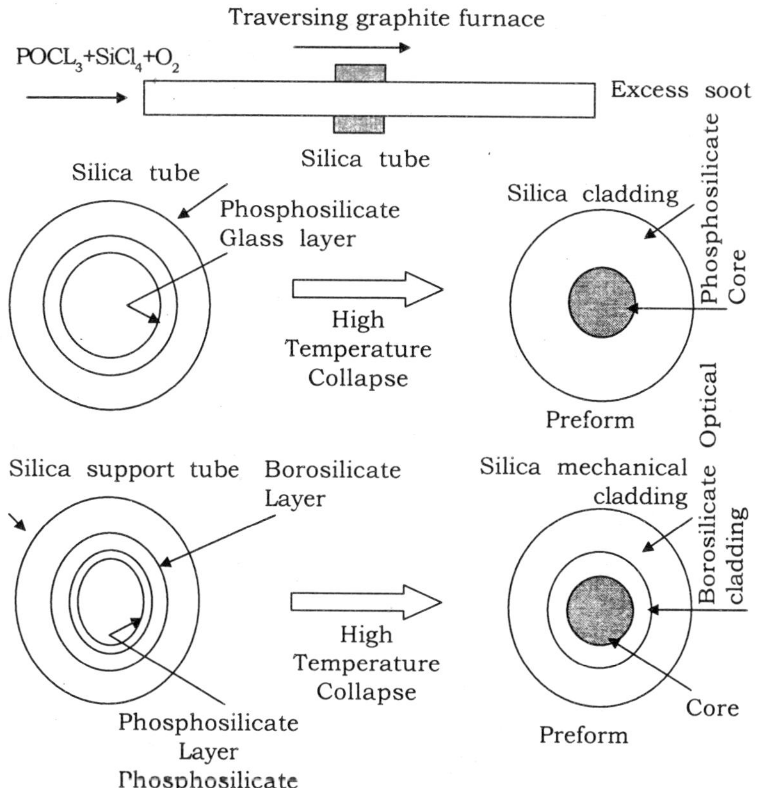

Fig.2.1: Two-layer process of manufacture of phosphosilicate fibers, with silica tube acting as a supporting structure.

The typical diameter lies in the range of 25-60 mm and 600-=800 mm in length.

The broadening of the light pulses occurs as they are transmitted through the fiber producing inter symbol interference (ISI) (a) fiber input, (b) the output pulse of the fiber at distance L_1 (c) fiber output at a distance L_2 where $L_2 > L_1$.

There are three kinds of dispersion due to three separate mechanisms existing in the fiber. These are:

(1) Inter modal Dispersion or Multipath dispersion

(2) Material or chromatic dispersion and

(3) waveguide dispersion.

INTERMODAL OR MULTIPATH DISPESION

The effective group velocity of light wave is different for each mode that wave guide will support, even though the phase velocity in different mode ray may be identical. This happens because total path followed by a ray, in each of the mode is zigzag in nature and the path in a particular mode has different total length than each of the other mode rays.

When a pulse of light is transmitted in a fiber, it will be propagated over several different paths corresponding to the excited modes and come out of the far end at slightly different times. The received pulses at the far end is the summation of these mode pulses each delayed by different amount of time. The shortest time delay will be suffered by the plane wave propagating straight along the axis of the fiber core: the longest time delay will be suffered by the wave propagating with the highest order supported mode where the angle of incidence is slightly greater than the critical angle θ_c. The difference in the arrival time Δt (propagation time delay) can easily be calculated for a step index fiber [2].

Let us assume that the total fiber length is Z. It is convenient to express Z in kilometres. Figure 2.2 shows two rays of different modes that are moving in zig zag paths along the length of the fiber. The length Z_t of the zigzag path is found to be

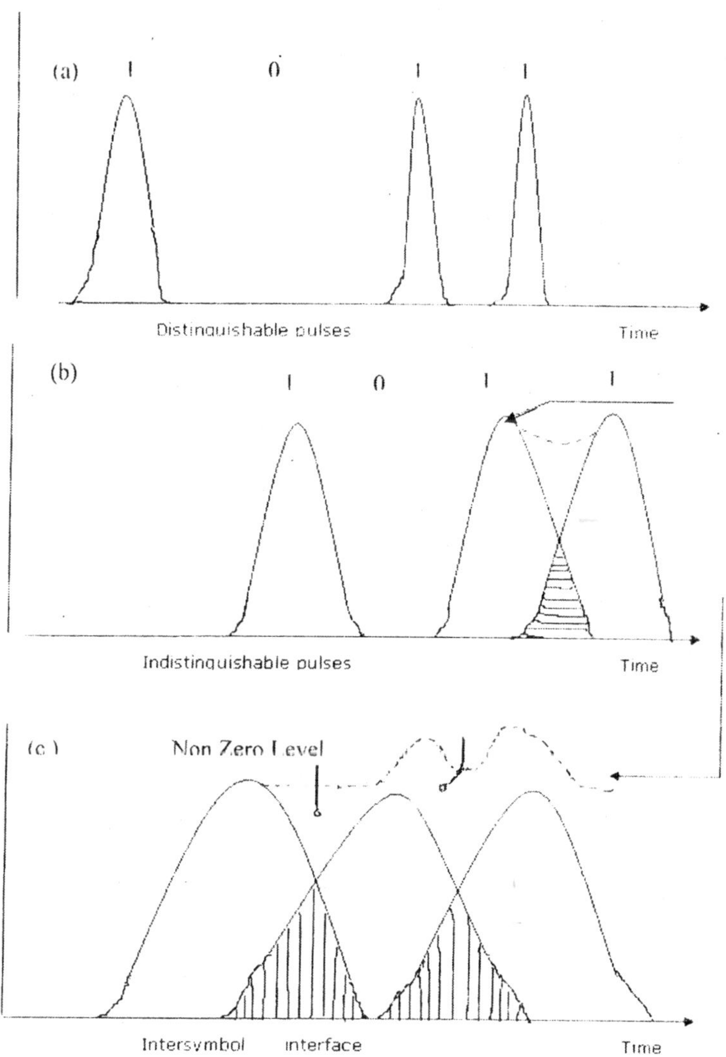

Fig. 2.2: An illustration of Pulse Broadening

$$Z_t = \frac{Z}{\text{Sin}\phi} \tag{2.1}$$

When ϕ is the angle of incidence in the mode with the normal to the fiber wall In the lowest order mode, $\phi = \phi_{max} = 90^0$

Thus, the minimum path length

$$Z_t = \frac{Z}{\sin\phi_{max}} = \frac{Z}{\sin 90°} = Z \tag{2.2}$$

and ϕ_{min} the longest path is found to be maximum path length,

$$Z_{max} = \frac{Z}{\sin\phi_{max}} = \frac{Z}{\sin\theta_c} = Z\frac{n_1}{n_2} \tag{2.3}$$

The maximum path difference ΔZ in the path due to modal separation is found to be

$$\Delta Z = Z_{max} - Z_{t\,min} = Z\left[\frac{n_1}{n_2} - 1\right] = Z\left[\frac{n_1 - n_2}{n_2}\right] \tag{2.4}$$

The maximum path difference depends on the difference of refractive index between the core and the cladding. Since the light ray in the fiber is propagating through a dielectric medium with a dielectric constant which is greater than unity, the ray travels more slowly than that they would move in free space. It is assumed that the relative permeability of the dielectric medium is unity. The phase velocity in the dielectric along the ray path is shown in Fig. 2.3.

Phase velocity

$$V_p = \frac{1}{\sqrt{\mu\varepsilon}} = \frac{1}{\sqrt{\mu_0\varepsilon_0}} = \frac{1}{\sqrt{\mu\varepsilon}} = \frac{C}{\sqrt{\varepsilon_r}} = (\mu = 1) \tag{2.5}$$

Since

$C = \frac{1}{\sqrt{\mu_0 \varepsilon_0}}$ = velocity of light in free space

= 3×10^8 m/s thus the dielectric constant ε_r is

$$\sqrt{\varepsilon_r} = \frac{C}{V_p} = n = \text{refractive index of the medium} \quad (2.6)$$

Thus, we find that the relationship between the dielectric refractive index n_1 is given by

$$\varepsilon_r = n^2 \quad (2.7)$$

The phase velocity in the glass core of the fiber with refractive index n_1 is given by

$$\text{Phase velocity } V_{p(glass)} = \frac{C}{n_1} \quad (2.8)$$

Finally, the maximum difference in the time delay between the rays of highest and lowest order modes is obtained by dividing the difference in the path length by the phase velocity. Denoting this by t and is given by Fig 2.3:

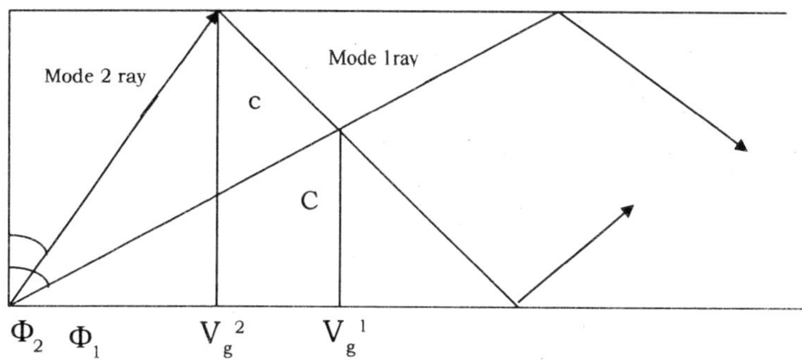

Fig 2.3 Two rays of different modes re moving in a zigzag paths along a fiber

$$\Delta t = \frac{\Delta z}{V_p} = \frac{n_1 Z}{C}\left[\frac{n_1 - n_2}{n_2}\right] = \left[\frac{Z}{C}\right]\left[\frac{n_1(n_1 - n_2)}{n_2}\right] \tag{2.9}$$

Denoting the difference between the refractive indices at the core and cladding (n_1- n_2) by Δn, the maximum time difference between becomes

$$\Delta t = \frac{Z}{C}\frac{n_1}{n_2}\Delta n \quad \text{or} \quad \frac{\Delta t}{Z} = \frac{n_1}{n_2 C}\Delta n \tag{2.10}$$

The term $\Delta t/Z$ is known as the "multiple time difference" of the fiber.

MATERIAL DISPERSION

The refractive index of core glass of fiber is not the same for lights of different wavelengths but it depends on the wavelength of light in a complicated way. When a pulse of light transmitted contains components of several wavelengths centered about a center wavelength as shown in Fig 2.4. The pulse component having

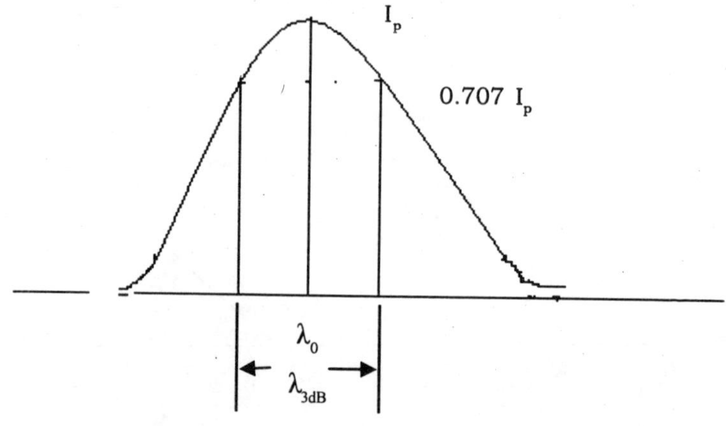

Fig 2.4: Defining the spectral width of the light sources: Central wavelength λ_0:$2\lambda_{3dB}$ = Δ

shorter wavelengths will experience move delay that those of larger wavelengths. As a result there is an effective time dispersion of the pulse at the receiving end of the fiber [3-5].

RELATIONSHIP BETWEEN THE MATERIAL DISPERSION AND WAVELENGTHS

Material dispersion is due to the explicit dependence of the refractive index of the core cladding on the wavelength (λ_o). It is a characteristic of the material only. Material dispersion plays a very important role in design of a fiber optics communication system. When a temporal pulse passes through a homogeneous medium, it propagates with a group velocity (V_g) given by

$$1/V_g = dk/dw$$

$$\Rightarrow \frac{1}{V_g} = \frac{d}{dw}(K)$$

$$= \frac{d}{dw}\left[\frac{n}{c} n(w)\right] \qquad (2.11)$$

Where $k(w) = \frac{n}{c}$ n(w), n(w) frequency–dependent refractive index.

In terms of wavelength, the group velocity becomes

$$\frac{1}{V_g} = \frac{1}{c}\left[n(\lambda_0) - (\lambda_0)\ \frac{dn}{d\lambda_0}\right] \qquad (2.12)$$

Thus, the time taken by a pulse to traverse length L of fiber is given by

$$\tau = \tau(\lambda_0) = \frac{L}{V_g} = \frac{L}{C}\left[n(\lambda_0) - (\lambda_0)\,\frac{dn}{d\lambda_0}\right]$$

Hence, temporal broadening of the pulse is

$$\nabla\tau = \frac{d\tau}{d\lambda_0} = \nabla\lambda_0 = \frac{L}{C}\left[\lambda_0^2\,\frac{d^2n}{d\lambda_0}\right]\left[\frac{\nabla\lambda_0}{\lambda_0}\right] \qquad (2.13)$$

The above broadening is referred to as material dispersion. Material dispersion in terms of ps/km-nm.

$$D_m = \frac{1}{\lambda_0 c}\left[\lambda_0^2\,\frac{d^2n}{d\lambda_0}\right]\left[\frac{\nabla\lambda_0}{\lambda_0}\right]10^9 \text{ ps/km-nm.} \qquad (2.14)$$

Where λ_0 is measured in micro meters.

COMBINED EFFECT OF MATERIAL DISPERSION AND INTERMODAL DISPERSION

In estimating the bandwidth and maximum bit rate of an optical communication channel we need to consider the shape and width of the receive pulse. The shape on the arrival of an impulse that has been

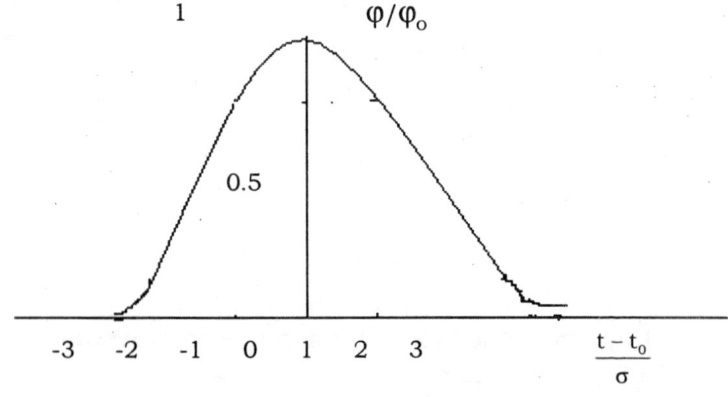

Fig 2.4(a) Gaussian distribution curve

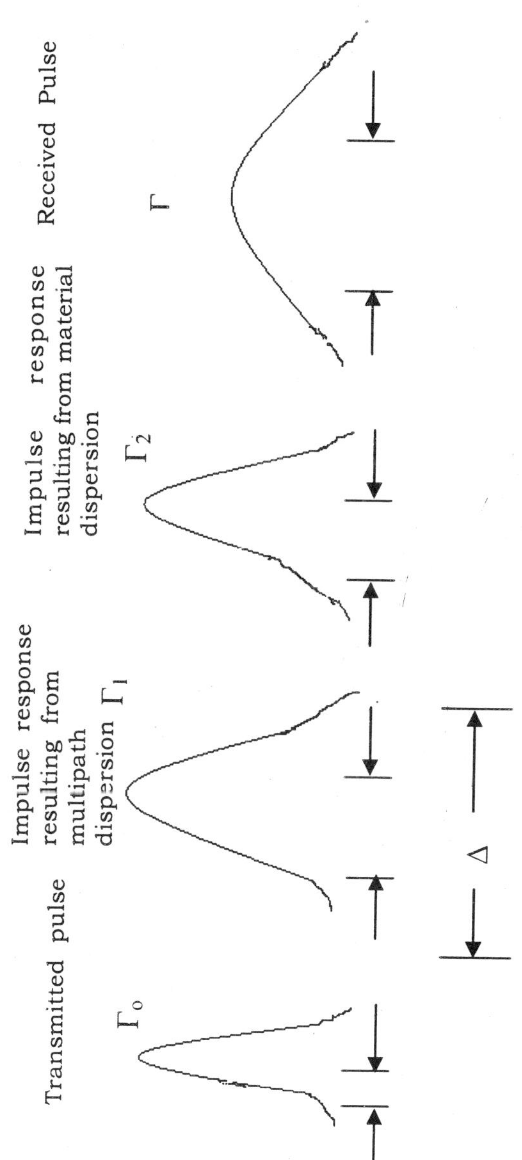

Fig. 2.4(b): Constant Phase Wave Front

broadened by material dispersion will control the power distribution among the wavelengths making up the pulse. We shall find in general that most optical sources have distribution of power with wavelengths that are roughly Gaussian in nature. We should then expect the receive pulse shape to take on a similar Gaussian form about the mean arrival time t as shown if Fig 2.4(a), (b). Now let us assume that the impulse is broadened by both material and multi path dispersions, and that the two mechanisms are uncorrelated and they independently lead to approximately Gaussian pulses having their half height pulse width G_1 and G_2 respectively. The two mechanisms will then combine to produce pulse which remains roughly Gaussian in shape and which will have a half width which is given by

$$\Gamma = \sqrt{\Gamma_1^2 + \Gamma_2^2} \tag{2.15}$$

When the transmitted pulse is not an impulse to start with but is also a Gaussian with a full width at the half height value of $\Gamma_{0,}$ then we may extend this argument as illustrated in Fig 2.5 to suggest that Γ becomes

$$v = \frac{w}{c} a\sqrt{n_1^2 - n_2^2} \; \Gamma = \sqrt{\Gamma_0^2 + \Gamma_1^2 + \Gamma_2^2} \tag{2.16}$$

where Γ_0 is the width, Γ_1 is the half height pulse width that would result from multi path broadening alone and Γ_2 represents the broadening due to material dispersion alone.

WAVE GUIDE DISPERSION

When a fiber could be so operated that the multipath and material dispersions are all eliminated as should be the case for a single mode fiber operation near λ = 1.3 mm, then a third dispersion mechanism will predominate except in the case ideally monochromatic light. Unfortunately ideal monochromatic courses are not available so that the light transmitted consists of component of several wavelengths near the central wavelength λ_0 within an envelope like that in Fig 2.5 . The result is that the constant phase wave front separation as illustrated in Fig 2.5 will be slightly different for different wavelengths and the corresponding angle of incidence will also be slightly different. This will cause a corresponding shift in the group velocity. The group velocity varies in complicated way with the wavelength and the net result is the broadening of the received pulse. This effect is only the result of the guiding characteristics of the fiber and hence it is known as the "wave guide dispersion".

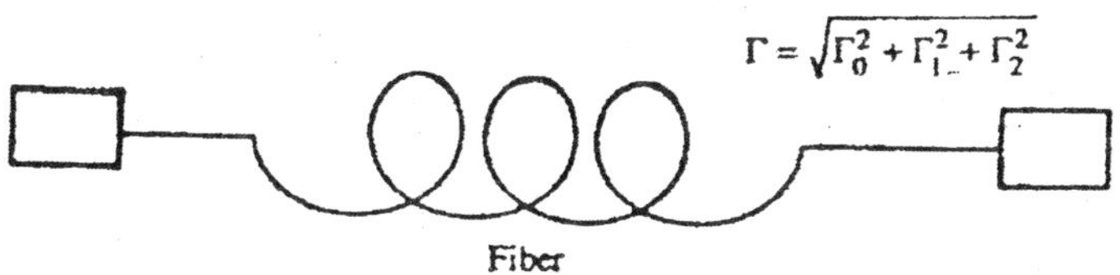

Fig 2.5: Dispersive effect in combination

WAVE GUIDE DISPERSION - EQUATIONS

Wave guide dispersion is due to the explicit dependence of the refractive index of the core and cladding on the frequency (w). In case of wave guide dispersion, wave propagation vector

$$\beta = \frac{w}{c}[n_2 + (n_1 - n_2)b(v)] \tag{2.17}$$

where *b*-normalized propagation constant,

$$v = \frac{w}{c} a\sqrt{n_1^2 - n_2^2} \qquad (2.18)$$

The waveguide parameter, a is the radius of the core, n_1 and n_2 are the core and cladding refractive index.

Hence, group velocity is given by

$$\frac{1}{v_g} = \frac{d\beta}{dw} = \frac{n_2}{c} + \frac{n_1 - n_2}{c}\left[\frac{d}{dv}(bv)\right] \int\langle\ \rangle \qquad (2.19)$$

Thus, the time taken by a pulse to traverse length *L* of the fiber is given by

$$\tau_w \frac{L}{v_g} = \frac{L}{c} n_2 \left[1 + \Delta \frac{d}{dv}(bv)\right] \qquad (2.20)$$

where $\Delta = \dfrac{n_1^2 - n_2^2}{2n_1^2} \cong \dfrac{n_1 - n_2}{n_2}$ (2.21)

For a source having a spectral width $\Delta\lambda_0$, the corresponding waveguide dispersion is given by

$$\Delta\tau_w \frac{d\tau}{d\lambda} \Delta\lambda_0 \approx \frac{L}{c} n_2 \Delta \left[\frac{\Delta\lambda_0}{\lambda_0}\right] v \left[\frac{d^2(bv)}{dv^2}\right] \qquad (2.22)$$

In terms of ps/km-nm, the waveguide dispersion is given by

$$D_w \frac{n_2 \Delta}{c\lambda_0} \left[v \frac{d^2(bv)}{dv^2}\right] 10^7 \text{ ps/km.nm} \qquad (2.23)$$

Where λ_0

$$\left[v \frac{d^2(bv)}{dv^2}\right] = \frac{2B^2}{dv^2} \qquad (2.24)$$

For a step index fiber the dispersion coefficient resulting from wave guide dispersion mechanism has a peak value of about 0.6 ps/nm. The value is an ideal lower limit on the dispersion at the wavelength where the material dispersion disappears. A partial cancellation does take place between the material dispersion and wave guide dispersion which ahs a zero cross point as shown in Fig. 3.18 slightly higher wavelength. By turning the light source and keeping the bandwidth now it is possible to remain practical value of loss less than 6.6. ps/km.

EFFECT OF NON-LINEARITY IN DISPERSION COMPENSATION FIBER

Here it has been treated that the optical fiber as a nonlinear medium, that is intensity dependent refractive index. Consider the propagation of an optical pulse in a loss fiber characterized by the refractive index.

$$n(w)=n_o(w)+n_{20}I, \qquad (2.25)$$

where $n_o(w)$ is the linear part of the refracting index, $n_{20}I$ is the intensity-dependent nonlinear part.

In case of material dispersion, if it use nonlinear effect, the group velocity (v_g) is given by

$$\frac{1}{v_g}=\frac{dk}{dw}=\frac{d}{dw}\left[\frac{w}{c}n(w)\right], \qquad (2.26)$$

where $k=\frac{w}{c}n(w)$

$$\frac{1}{v_g}=\frac{d}{dw}\left[\frac{w}{c}n_o(w)+n_{20}I\}\right],$$

$$\frac{1}{v_g} = \frac{1}{c}\left[n_o(w) + n_{20}I + w\left(\frac{dn_o}{dw} + I\frac{dn_{20}}{dw} \right) \right]. \tag{2.27}$$

Now, group velocity in terms of the free space wavelength $\lambda_0 (= 2\pi c / w)$ is given by

$$\frac{1}{v_g} = \frac{1}{c}\left[n_o(\lambda_0) + n_{20}I + \lambda_0 \frac{dn_0}{d\lambda_0} - \lambda_0 \frac{dn_{20}}{d\lambda_0} \right]. \tag{2.28}$$

Hence, the time taken by the pulse to traverse length L of the fiber is given by

$$\tau = \tau(\lambda_0) = \frac{L}{v_g}\left[n_o(\lambda_0) + n_{20}I + \lambda_0 \frac{dn_0}{d\lambda_0} - \lambda_0 \frac{dn_{20}}{d\lambda_0} \right]. \tag{2.29}$$

If the spectral width of the source is $\nabla\lambda_0$, the temporal broadening of the pulse is given by

$$\nabla\tau = \frac{d\tau}{d\lambda_0}\nabla\lambda_0 = -\frac{L}{c}\left[\lambda_0 \frac{d^2 n_0}{d\lambda_0^2} + \lambda_0 I \frac{d^2 n_{20}}{d\lambda_0^2} \right]$$

$$= -\frac{L}{c}\left[\lambda_0^2 \left\{ \frac{d^2 n_0}{d\lambda_0^2} + \frac{d^2 n_{20}}{d\lambda_0^2} I \right\} \left(\frac{\nabla\lambda_0}{\lambda_0} \right) \right] \tag{2.30}$$

We know $n_{20} = \dfrac{3\chi^3}{4c \in_0 n_0^2}$

$$\frac{d^2 n_{20}}{d\lambda_0^2} = + \frac{9\chi^3}{2c \in_0 n_0^4} \frac{d^2 n_0}{d\lambda_0^2} = \frac{6n_{20}}{n_0^2} \tag{2.31}$$

Thus, $$\nabla\tau_0 = -\frac{L}{c}\left[\left(\lambda_0^2 \frac{d^2 n_0}{d\lambda_0^2} \right) \left\{ 1 + \frac{6n_{20}}{n_0^2} I \right\} \right] \left(\frac{\nabla\lambda_0}{\lambda_0} \right), \tag{2.32}$$

$$D_{mn} = -\frac{10^9}{\lambda_0 c}\left[\left(\lambda_0^2 \frac{d^2 n_0}{d\lambda_0^2}\right)\left\{1+\frac{6n_{20}}{n_0^2} I\right\}\right], ps/km-nm. \quad (2.33)$$

In case of waveguide dispersion, it is considered that v (waveguide parameter) is approximately constant. Then

$$D_{mn} = -\frac{n_2}{n_2}\left(n_0 + n_{20} I - n_2\right)\frac{10^2}{3\lambda_0}\left(\frac{d^2(bv)}{dv^2}\right), \quad (2.34)$$

$$D_{mn} = -\frac{n_2 \nabla}{c\lambda_0}\left(v\frac{d^2(bv)}{dv^2}\right)10^2\left(1+\frac{n_{20} I}{\nabla}\right) ps/km-nm, (iii)$$

(2.35)

where $\nabla = \frac{n_0 - n_2}{n_2}$

POLARIZATION MODE DISPERSION

Polarization doesn't matter in multimode fibers, but it can be important in single mode fibers. The reason is that single mode fibers actually catty two modes with orthogonal polarizations. From the waveguide theory, the two modes with degenerate, i.e., they are functionally identical, so light can shift easily between the two polarization modes. Polarizations have little practical impact for communication, if the circular symmetry of the fiber is perfect. In addition, the forces affecting the fiber are never applied symmetrically around the fiber. Due to this fact, the two polarization modes experience different travel conditions with slightly different speeds resulting in differential group delay. The PMD was measured using interferometric method[2, 11]. In principle, it measures the differential group delay and principal state of polarization as a function of frequency. In our work, fiber under test was coupled test was coupled between transmitter and receiver. The collimated light from a broad band light emitting diode was polarized and split into two mutually coherent beams and passed over two mirrors. One mirror can be scanned in a position, creating a differential delay between the two orthogonal polarizations, which were recombined and directed through the fiber under test. When photocurrent fringes can be observed only when this differential delay was compensated by the differential group delay of the fiber under test. This measurement is closely related to the direct measurement in the time domain: it essentially measures the autocorrelation of the out coming "light

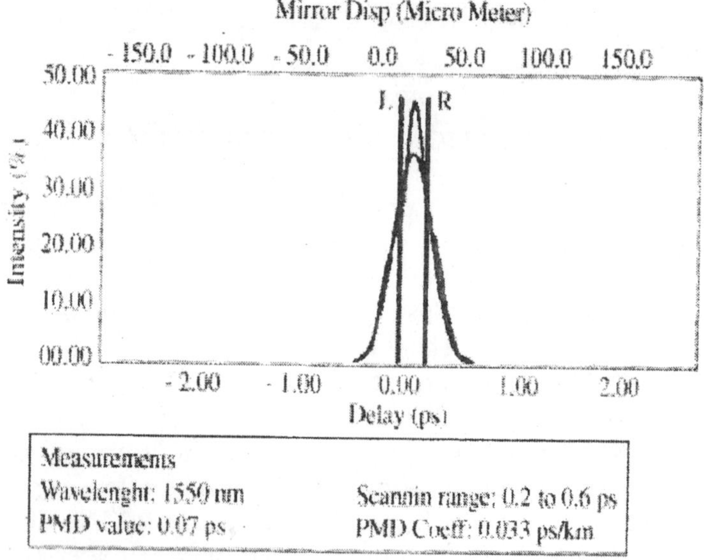

Fig. 2.6: The Differential Group Delay in PMD using Interferometric Method.

pulse". The coherence time of the light sources is playing the role of the pulse width. This coherence time determines thus the smallest measurable polarization mode delay. Figure 2.6 shows the plot of differential group delay observed due to polarization mode dispersion. The central peak occurs when the interferometer arms are of equal lengths. The outlying occurs when the movable mirror introduces a delay equal to the differential group delay of the test device. L peak is due to interference between fast mode light originating in the fixed mirror path and fast mode light delayed by the movable mirror. The R peak is due to interference between fast mode light originating in the fixed mirror path and slow mode light advanced by an identical movement of the mirror in the opposite direction. The time separation of the central and either of the outer peaks L and R, is the differential group delay of the device. PMD in relation to the coherence time of the source was noted as 0.07 ps at 1550 nm.

REFERENCES

1. John M.Senior, *Optical Fiber Communications,* Printice-Hall International (UK) Ltd. 1992.
2. Cristani, *Opt: Communn.*, Vol.146, 1998
3. S.K.Turistyn and E.G.Shapiro. *Opt. Lett* 23, 1998.
4. K. Prozean, *Proc. Indian Nat. Sci Acad.* 67, 2001.
5. D. Anderson, *Proc. Inst.Electr.Engg.* Vol. 123, 1985
6. Heffner, B.L., Hernday, R., Measurement of polarization-mode dispersion. *Journal of Hewlett-Packard* 1995; 46(1): 27-33.
7. Obserson, P., Juliard, K., Gisin, N., Passy, R. and Weid, J., Interferometric polarization mode dispersion measurement with femtoseconds sensitivity. *Journal of Light wave Technology.* 1997; 15(10): 1852-1857.

3

Techniques to Measure Polarization Mode Dispersion

INTRODUCTION

Since the introduction of the early concepts (1, 2), the fundamentals of polarization mode dispersion (PMD) in optical fibers have become an important body of knowledge basic for the design of high-capacity optical communication systems. Polarization mode dispersion (PMD) is a form of model dispersion where two different polarizations of light in a waveguide, which normally travel at the same speed, travel at different speeds due to random imperfections and asymmetries, causing random spreading of optical pulses. Unless it is compensated, which is difficult, this ultimately limits the rate at which data can be transmitted over a fiber.

Polarization mode dispersion effects are linear electromagnetic propagation phenomena occurring in so-called "single-mode" fibers. Despite their name, these fibers support two modes of propagation distinguished by their polarization. Because of optical birefringence in the fiber, the two modes travel with different group velocities, and the random change of this birefringence along the fiber length results in random coupling between the modes. With current practical transmission technology the resulting PMD phenomena lead to pulse distortion and system impairments that limit the transmission capacity of the fiber. Excellent reviews are available (3, 4), covering the practical aspects and applications of PMD concepts to fiber transmission systems and the effects of PMD on nonlinear fiber transmission (5). In this review we aim to complement these surveys and to collect and synthesize the fundamental concepts and theory of PMD, interweaving and linking the principal laws and key formulas that appear scattered in various places in the literature.

We examine the PMD vectors and principal states of polarization (PSPs) used to describe light propagation in randomly birefringent fibers and the associated differential group delay (DGD). Many different techniques for PMD measurement are available, and often many user-selectable parameters are associated with the measurement. This presentation will describe the various measurement methods and discuss the best practices for measurements with each method. Increasing consumer demand means that there is a requirement for higher rates of data transmission through fibre optic systems. Current capabilities are approximately 10 Gbits/s, and one of the limiting factors preventing higher rates is Polarization Mode Dispersion.

The PMD can be present in both optical fiber and fiber optic components, but 'investigations have previously been focused on optical fiber or systems as a whole. Now, as the contribution of Single Mode (SM) fiber to the system's PMD has been minimized through improved technology, PMD contributions from components are becoming more important. Components can be considered to be devices such as fiber Bragg gratings, isolators, circulators and Dense Wavelength Division Multiplexing (DWDM) devices. It is of particular interest to be able to quantify the PMD of DWDM devices as they present a way of increasing data rates using existing infrastructure.

The requirement is that devices manufactured for a system do not intolerably increase the system's PMD. As such manufacturers, including SMEs, are being given specifications for the PMD of their products, and so there is a demand for measurement of PMD in components. As current techniques used were developed for the measurement of PMD in single mode fiber, their suitability for application to components needs to be considered. This report does this through literature-based investigations. A figure of merit, FOM, will be assigned to each technique as an indicator of its fitness for purpose to measure component PMD. It is calculated from:

$$\text{Figure of Merit} = \frac{1}{\text{DGD [s]} \times \text{bandwidth [Hz]}} \qquad (3.1)$$

where, 'bandwidth' is the spectral range needed to measure the associated DGD, Differential Group Delay. The higher the figure of merit the more suitable the technique.

Physically, Polarization Mode dispersion is the dispersion, or broadening, of a light signal travelling through an optical material due to the presence of two different refractive indices (birefringence [6]), which transmit polarized light at two different speeds. Birefringence in components generally stems from either an element within the component that is intrinsically birefringent, or a piece of fiber within a device that is exhibiting birefringence [7]. There are many mechanisms that cause birefringence in fiber.

The simple action of applying stress to, or tightly bending a single mode fiber induces birefringence, and Highly Birefringent (Hi-Bi) or Polarization Maintaining (PM) fiber may have been engineered through geometrical asymmetries, for example. In fiber, birefringence is manifested as two optical axes known as the fast and slow principal axes. Each axis has a Principal State of Polarization (PSP) [8] associated with it, and these two polarization states are orthogonal when no Polarization Dependent Loss (polarization sensitivity) is present.

When light is launched fully into a PSP, the output state is invariant for different wavelengths to first order, and the input state propagates unchanged provided no mode coupling occurs. Mode coupling occurs at a change in birefringence or at a junction between birefringent elements. These concepts, along with Differential Group Delay, DGD, provide a physical basis for understanding PMD. DGD is the differential delay in arrival times between light from the PSPs. DGD is the maximum delay when there is no Polarizations Dependent Loss, but is not necessarily the maximum delay when the PSPs are not orthogonal [9, 10]. PMD is defined as an average of DGD values over a wavelength range, and can therefore be thought of as a statistical quantity. The main purpose of this chapter is to show brief theoretical backgrounds of the *PMD* in single mode optical fibers and the principles of the utilization of polarization optical time-domain reflectometry (PO-OTDR) for *PMD* measurement. There are several methods frequently used for the *PMD* measurements.

CAUSES OF PMD

The major cause of PMD is the asymmetry of the fiber-optic strand. Asymmetry is simply the fact that the fiber core is slightly out-of-round, or oval given in Fig. 3.1. Fiber asymmetry may be inherent in the fiber from the manufacturing process, or it may be a result of mechanical stress on the deployed fiber. The inherent asymmetries of the fiber are fairly constant over time, while the mechanical stress due to movement of the fiber can vary, resulting in a dynamic aspect to PMD.

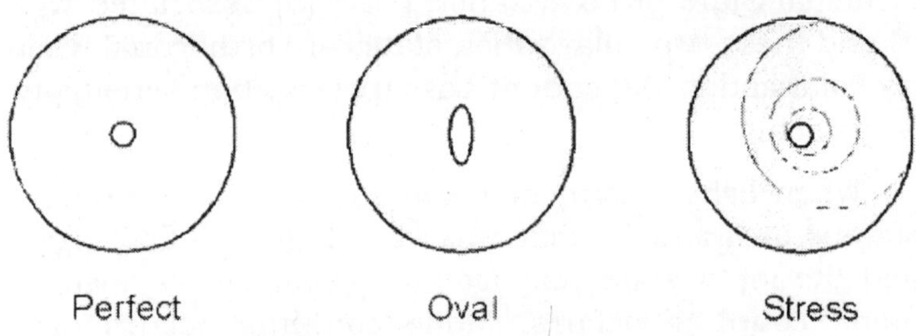

Fig. 3.1: Cross-Sections of Optical Fibers

The mechanical stress on the optical fiber can originate from a variety of sources. One source that is very difficult to control is the diurnal (day/night) and seasonal heating and cooling of the optical fiber. Although much fiber is deployed in the ground and often within conduits, it is still subject to temperature variations and corresponding mechanical stress.

Another source of mechanical stress can originate from nearby sources of vibration. For example, much fiber is deployed alongside railroad tracks because of the ease of right-of-way and construction. However, vibration from passing trains can contribute to stress on the optical fiber. Fiber that is not buried next to railways and highways may be deployed aerially; in this scenario, wind can cause swaying of the fiber cable and can contribute to PMD.

Because of the combination of these effects, and the random way these effects add up over a section of fiber,

PMD does not have a single value for a given section of fiber. Rather, it is described in terms of average DGD, and a fiber has a distribution of DGD values over time. The probability of the DGD of a fiber section being a certain value at any particular time follows a Maxwellian distribution given in Fig 3.2. As an approximation, the maximum instantaneous DGD is about 3.2 times the average DGD of a fiber.

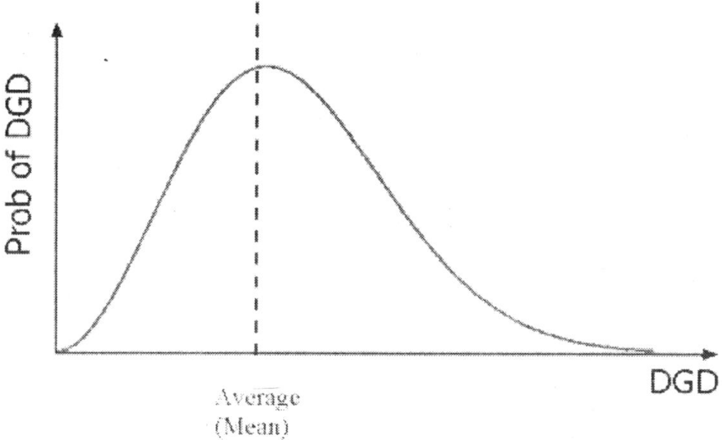

Fig. 3.2: Probability Distortion of DGD Levels in a Typical Fiber

THE PMD COMPENSATOR

A PMD compensation system is a device which uses a polarization controller to compensate for PMD in fibers. Essentially, one splits the output of the fiber into two principal polarizations (usually those with $d\tau$ / $d\varpi = 0$, i.e. no first-order variation of time-delay with frequency, and applies a differential delay to bring them back into synch. Because the PMD effects are random and time-dependent, this requires an active device that responds to feedback over time. Such systems are therefore expensive and complex; combined with the fact that

PMD is not yet the limiting factor in the lower data rates still in common use, this means that PMD-compensation systems have seen limited deployment in large scale telecommunications systems. Another alternative would be to use a polarization maintaining fiber PM fiber, a fiber whose symmetry is so strongly broken that an input polarization along a principal axis is maintained all the way to the output. Since the second polarization is never excited, PMD does not occur. Such fibers currently have practical problems, however, such as higher losses than ordinary optical fiber and higher cost. An extension of this idea is a single polarization fiber in which only a single polarization state is allowed to propagate along the fiber.

The PMD compensation, like regeneration, is performed on a per-wavelength basis. This is due to significant variability in PMD over wavelength within a transmission band. However, a single PMD compensator at the receive side of a 10 Gbps signal can mitigate the PMD effects that build up over an entire transmission span. Therefore, unlike the regeneration option that may require two separate sites to eliminate PMD on a 600 km span, a single PMD compensator located at the destination or receive side of the optical signal can improve the end-to-end performance of the transmission path. PMD compensation can be accomplished for a fraction of the cost of a single regenerator site's equipment costs. In addition to the equipment costs associated with regenerating optical signals, there is also the issue of the costs of the regenerator sites themselves. Since these regenerator sites must be located within certain distance ranges along the optical transmission path in order to operate most effectively, the required placement for these sites may often be in difficult geographic locations. The buildings to accommodate these sites are often more costly to construct and maintain than equivalent real-estate space in the most desirable urban locations, simply because the site is so remote.

There are several technologies that can be utilized to compensate for the effects of PMD. These technologies have included the following:

(a) Mechanical devices that actually squeeze a portion of the fiber in order to realign the polarization pulses of the optical bit. In other words, a mechanical PMDC "counter-stresses" the fiber. The primary drawback of this method is that mechanical devices are more prone to failure over long durations; e.g., the original mechanical "step" switches of early voice networks, a technology of the early 20th century, demanded frequent maintenance and were prone to problems.

(b) Electronic devices that work after the receiver decoder, manipulating electrons in order to reduce bit errors. The primary drawback to this method is the difficulty in correcting an optical problem at the electronic layer.

The most reliable and efficient PMDC technology is the use of adaptive optics to realign and correct the pulses of dispersed optical bits A related effect is polarization-dependent loss (PDL), in which two polarizations suffer different rates of loss in the fiber due, again, to asymmetries. PDL similarly degrades signal quality.

Strictly speaking, a circular core is not required in order to have two degenerate polarization states. Rather, one requires a core whose symmetry group admits a two-dimensional irreducible representation. For example, a square or equilateral-triangle core would also have two equal-speed polarization solutions for the fundamental mode; such general shapes also arise in photonic crystal fiber. Again, any random imperfections that break the symmetry would lead to PMD in such a waveguide.

TWO DIFFERENT MODES OF PMD

One can distinguish two of appropriate meaning *PMD* definitions. The first definition is analogous to the definition of modal dispersion for multimode fibers [11]. This is not surprising because the single-mode optical fiber actually supports several modes called "polarization modes" (PM). In the absence of PM coupling one has only two modes. But in the more realistic case of random PM coupling the fiber behaves as multi-mode fiber and consequently the natural definition for polarization mode delay (PMD_1) is the mean square deviation of the time of flight "t" of various PM:

$$PMD_1 = \Delta\tau = 2[(t^2) - (t)^2]^{1/2} = 2\left[\frac{\int I(t)t^2dt}{\int I(t)dt} - \left(\frac{\int I(t)tdt}{\int I(t)dt}\right)^2\right]^{1/2} \quad 3.2$$

Here $I(t)$ is the intensity of the light pulse after propagating through the fiber assuming the initial pulse of the $\delta(t)$ shape. It can be shown that in the case of high polarization mode coupling $I(t)$ is of the Gaussian form. The second definition is suitable for the measurement on systems where the coherence time of the optical source is larger than the PM delay. In that case for fixed PM coupling and the fixed wavelength each polarization state has a fixed group velocity. The fastest and slowest polarization states are called the principal states of polarization (PSP). For these states the difference between their group velocities are very sensitive to the wavelength and to the PM couplings.

It is suitable to call instantaneous *PMD* the delay between the principal states and to consider their average. Consequently the second definition is

$$PMD_2 = \tau(\omega)_\omega = \frac{\int_{\omega_1}^{\omega_2} \tau(\omega) d\omega}{\omega_1 - \omega_2} \tag{3.3}$$

These two "different" definitions describe the same physical quantity so it should be possible to relate them. The relation between the two definitions can be found using the basic formula [12]

$$PMD_1 = \Delta\tau = \left[\frac{\int_{\omega_1}^{\omega_2} \tau(\omega)^2 d\omega}{\omega_2 - \omega_1} \right]^{1/2} \tag{3.4}$$

which is valid for any degree of PM coupling.

Using this formula one can find that for low PM coupling the two definitions give the same results and for the high PM coupling $PMD_2 = PMD_1 / 1.085$ [13].

Where $L = v_g T_o / 2$. the above equations, can also be rewritten in a more compact form as follows

$$P_{bs\|bs\perp}(z) = \left[(P_{inx} \cos^2\gamma + P_{iny} \sin^2\gamma) \pm \sqrt{P_{inx} P_{iny}} \sin 2\gamma \left\{ \frac{\sin(\delta\beta v_g \frac{T_o}{2}}{\delta\beta v_g \frac{T_o}{2}} \right\} . \cos\delta\beta z \right]$$

$$S_{xy} \frac{v_g T_o}{2} a_r \exp(-2az) \tag{3.5}$$

The above relations reveal that the backward scattered power in the *PO-OTDR* contains the periodical fluctuation term determined by $2(\beta^{(\perp)} - \beta^{(\uparrow\uparrow)})z$ as well as the dc component which relates to the attenuation loss of the fiber (see the Figure 3.3). They also reveal that fluctuation components of $P_{s\uparrow\uparrow}$ and $P_{s\perp}$ are exactly out of the phase and the fluctuations disappear by adding $P_{s\uparrow\uparrow}$ and $P_{s\perp}$ Consequently the total back-scattered power does not fluctuate. Moreover the magnitude of the fluctuations increases with increasing the input optical power. As it can be seen in Eq. (3.5). the amplitude of the fluctuation term depends also strongly on the "interference" function that is of the type

(sin x)/x,

where $x = (\delta\beta v_g T_o)/2$.

Consequently the amplitude of the fluctuation term will be higher for shorter pulses and for a suitable choice of T_o it can be equal to zero. But this conclusion holds exactly only for the rectangular shape of the excitation impulse.

For high quality optical fibers the beat length of the fiber defined as $L_B = [2\pi/(\delta\beta)]$ falls into the region of several tens of meters [14, 15].

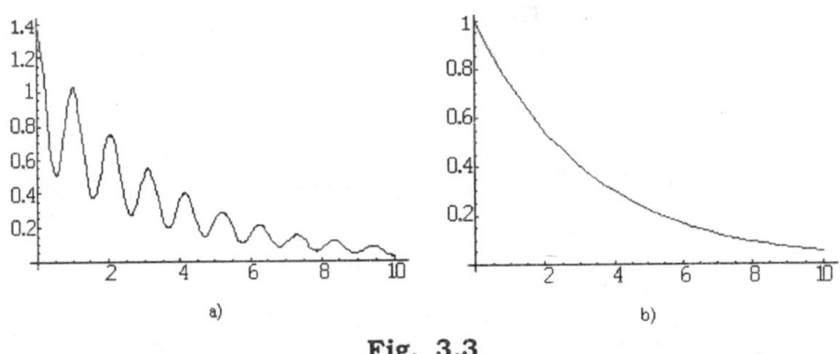

Fig. 3.3

Fig. 3.3(a) The qualitative representation of the fluctuation component of the back-scattered power, (b) The dc component of the back-scattered power detected without analyzer. For high response the pulse width can be T_0 = 12.5 ns the maximum measurable *PMD* by this setup was 4 ps/km and the length of the fiber should not exceed 10 km as shown in Fig.3.4.

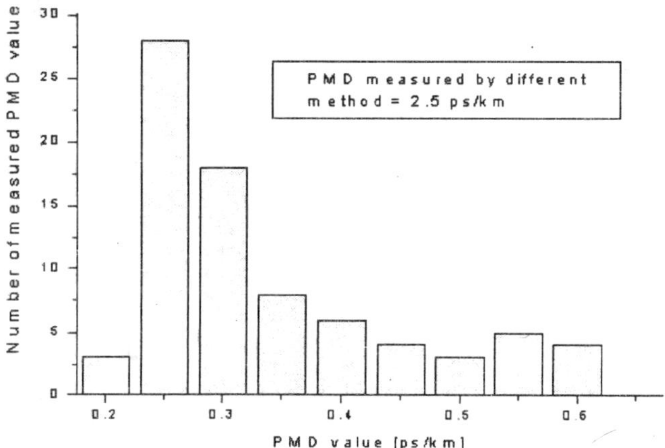

Fig.3.4: Histogram of *PMD* Values for Low Birefringence Single-mode Fiber

PMD MEASUREMENT TECHNIQUES

Measuring polarization-mode dispersion (PMD) can be confusing. Many different techniques for PMD measurement are available, and often many user-selectable parameters are associated with the measurement. This presentation will describe the various measurement methods and discuss the "best practices" for measurements with each method. The various measurement techniques can be classified as either time-domain or frequency-domain techniques. The clearest separation between the two is seen in the relationship between the coherence time *Tc* of the measurement light and Δτ. A technique is considered to be in the time domain if $Tc < \Delta\tau$, and in the frequency domain if $Tc > \Delta\tau$.

Time-domain Measurements

Time of Flight

The time-of-flight measurement is the most intuitive, so we will consider it first. Figure 3.5 illustrates a narrow pulses of light transmitted through a non mode-coupled device of DGD $\Delta\tau = \tau_f - \tau_s$. The propagation delay through the device will be either be τ_f (if the input pulse is polarized along the fast PSP), τ_s (if it is polarized along the slow PSP), or if the polarization state of the pulse is between the fast and slow PSP, the pulse will be broken into two pulses with delays τ_f and τ_s, with relative intensities weighted according to the orientation of the input polarization state with respect to the fast and slow PSP. So, in this intuitive measurement technique, short optical pulses are launched into a test device and detected at the output. For launch polarizations between the two PSP, two received pulses are seen separated in time by the mean DGD. This technique is well known [16,17,18], but is often impractical since the pulse width limits the temporal resolution.

Low-coherence Interferometry

A closely related but more practical approach to time domain measurement is through low coherence interferometry. A spectrally broad (low-coherence) source sends light through the test device and into an interferometer. As the moveable arm of the interferometer is translated, interference fringes are seen at the detector only if the time-delay difference between the two arms matches a delay generated in the test device to within the coherence time of the source. Ignoring possible coherence effects, this condition can be written simplistically as

$$\left|(\tau_{arm,1} - \tau_{arm,2}) - (\tau_i - \tau_j)\right| < T_c \qquad ....3.6$$

where $\tau_{arm,1}$ and $\tau_{arm,2}$ are the time delays associated with propagation along each arm of the interferometer, τ_i and τ_j are two possible propagation times experienced by light travelling along the ith and j^{th} polarization paths through the device. T_c is the coherence time of the source (e.g. for a Gaussian source of spectral width $\Delta\lambda$ and center wavelength λ, $T_C = 0.664 \cdot \lambda^2 / c\Delta\lambda$, c is the speed of light) [19]. The above equation gives an intuitive picture of the interferogram shape.

Low-coherence interferometry can be used to measure mode-coupled devices as well. In this case, τ_i and τ_j can take on 2^{N+1} different values (where N is the number of mode-coupling sites in the artifact). This yields a delay histogram with $2^{N+2} - 1$ peaks. However, the separation of adjacent peaks can easily be less than the coherence time of the source and so the peaks are not necessarily distinguishable. The resulting interferogram envelope comes from the coherent addition of the various delays. In highly mode-coupled devices, it is customary to characterize the root-mean-square (RMS) value of the DGD over the wavelength range of the measurement. The "second moment"

$$\sigma_M = \sqrt{\frac{\int I(t)t^2 dt}{\int I(t)dt}} \qquad (3.7)$$

of this Gaussian-shaped delay histogram will generally yield the RMS DGD value $\int I(t)$ is the amplitude of the delay histogram and t is the time component). The relationship between the two is [20]:

$$\langle \Delta\tau^2 \rangle^{1/2} \approx \sqrt{\frac{3}{4}}\, \sigma_M \qquad (3.8)$$

The evaluation of Equation should be performed with caution. There will be non ideal features of the delay histogram that make it deviate from a true Gaussian. The most significant features are the autocorrelation peak and the noise floor. The integration limits for Equations must be chosen carefully in order to exclude these unwanted features, but this must be done iteratively so as not to incur a bias in the opposite direction due to excluding real data. Examples of procedures to correctly extract $\langle\Delta\tau^2\rangle^{1/2}$ from a mode-coupled interferogram are provided in and [21]. PMD measurements using low-coherence interferometry have several advantages over other techniques. The measurements can be done quickly–the time required is essentially the travel time of the interferometer mirror over the desired scan range (a few seconds). Unlike most of the frequency- domain techniques, low-coherence interferometry does not require numerical comparison between data sampled at two different points in time. This makes interferometry less susceptible to dynamic changes in the measurement path (such as movement of the fiber leads). The biggest disadvantage of low-coherence interferometry is the need to use a spectrally broad source, which precludes measurements in a narrow pass band. Another aspect of PMD measurements made using low-coherence interferometry is the effect of multipath interference (MPI). Multiple reflections within the measurement path will cause delays in the signal transmission and show up on the delay histogram indistinguishable from delays due to PMD. While it may be useful to measure these MPI effects, the user should be aware that low-coherence interferometry is unable to distinguish between MPI and PMD.

Frequency-domain Measurements

PMD measurements based in the frequency domain measure the same DGD as time domain measurements but from a different perspective. The most common approach to DGD measurement in the frequency domain involves a differential method where the difference in propagation delay between light travelling on the fast and slow PSP is measured by assessing its effect on the polarization state of light as it exits the test device. The polarization dispersion vector Ω is related to the change in output polarization state **S** as [22].

$$\frac{d\mathbf{S}}{d\omega} = \Omega \times \mathbf{S} \tag{3.9}$$

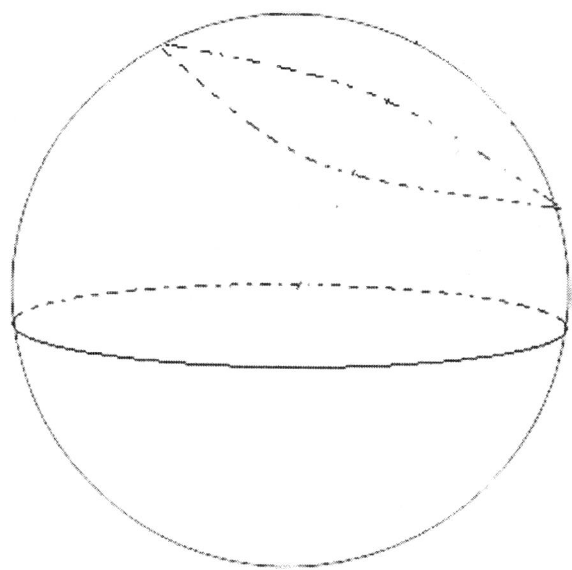

Fig. 3.5: Poincaré sphere representation of polarization dispersion vector Ω and output polarization state S at optical frequencies ω_1 and θ

where ω is the optical frequency of the light. The physical meaning of this expression is that PMD in a device causes the output polarization state to precess about the polarization dispersion vector (PSP) as the optical frequency is changed as described in Fig. 3.5.

From Equation (5), the precession rate will be equal to the DGD of the device:

$$\left|\frac{d\theta}{d\omega}\right| = |\Omega| = \Delta\tau \tag{3.10}$$

where θ is defined in Fig. 3.5 as the angle of rotation of the output state of polarization about the precession axis Ω. The DGD can be found by measuring $\Delta\theta/\Delta\omega$ (as an approximation to $d\theta/d\omega$). The class of frequency-domain techniques which measure $\Delta\theta/\Delta\omega$ will be referred to as "polarimetric techniques".

JME Technique

The JME technique gives a robust measurement of $\Delta\theta/\Delta\omega$ and Ω by turning the measurement into an eigen value problem [23]. At a given optical frequency ω_1, three noncollinear input polarization states such as linearly polarized with 0°, 45°, and 90° orientations are input to the device and the corresponding output polarization states are measured. This allows the 2 × 2 Jones transfer matrix $\boldsymbol{T}(\omega_1)$ [24]. ($\boldsymbol{T}(\omega_1)$ is the matrix that describes the transformation of the input polarization state to the output polarization state at the optical frequency

ω_1, $\boldsymbol{S}_{out}(\omega_1)=\boldsymbol{T}(\omega_1)\boldsymbol{S}_{in}$).

Then the same three states are launched at a slightly different optical frequency ω_2 and $\boldsymbol{T}(\omega_2)$ is calculated. The product

$\Gamma(\varpi, \Delta\omega) = \boldsymbol{T}(\omega_1)\boldsymbol{T}^{-1}(\omega_2)$,

where Γ, which depends on the average optical frequency $\varpi = (\omega_1+\omega_2)/2$, and on the frequency step size $\Delta\omega = \omega_2-\omega_1$, describes the evolution (precession) of the output polarization state about the PSP. That is,

$\boldsymbol{S}_{out}(\omega_2)=\Gamma(\varpi,\Delta\omega)\boldsymbol{S}_{out}(\omega_1)$.

The eigenvectors of G are the PSP, and the two eigenvalues are r*q*=exp(it$_{g',q}$Dw), where the index *q* denotes propagation along the fast or slow axis and t$_g$,q is the associated group delay. The DGD is then

$$\Delta\tau(\varpi) = \left|\tau_{g.f} - \tau_{g.s}\right| = \left|\frac{Arg(\rho_s)}{\Delta\omega}\right| \qquad (3.11)$$

Thus, JME allows the full Ω (ϖ) to be calculated, even for highly mode-coupled devices.

MME Technique

The Müller Matrix Method (MMM) [25] measures PMD in much the same way as the JME technique, but with a few simplifications. The MMM works entirely in Stokes space, with measurements of the output polarization state $\boldsymbol{S}$ and the calculation of the 3 × 3 polarization transfer matrix $\boldsymbol{R}$ all performed in Stokes space. JME typically measures $\boldsymbol{S}$ in Stokes space and calculates the 2 × 2 transfer matrix in Jones space. Second, the MMM technique measures the polarization transfer matrix by using only two launched states. These two states can simply be two different linear polarizations. The angle between them is not important to the measured value, but will have an effect on the measure-ment noise. Finally, this two-state characterization of the transfer matrix assumes no polarization-dependent loss (PDL) is present; JME does not make this explicit assumption. Similarly to JME, the MMM technique measures the polarization transfer matrix ***R*** at two closely spaced optical frequencies, yielding $\boldsymbol{R}(\omega_1)$ and $\boldsymbol{R}(\omega_2)$, and calculates $\mathbf{R}_{\Delta} = \mathbf{R}(\omega_2)\mathbf{R}\mathrm{T}(\omega_1)$. Since there is assumed to be no PDL, ***R*** and $\boldsymbol{R}_\Delta$ are treated as pure rotation matrices, meaning $\mathbf{R}^{T}=\mathbf{R}^{-1}$, so, under this assumption, $\boldsymbol{R}_\Delta$ is equivalent to Γ from the JME case. The difference is that $\boldsymbol{R}_\Delta$ is a 3 × 3 matrix in Stokes space and Γ is a 2×2 Jones matrix. If $\mathbf{R}_\Delta$ is a pure rotation matrix, the precession angle θ of the output polarization state will be given by

$$\cos(\theta) = \frac{1}{2}(TrR_\Delta - 1) \tag{3.12}$$

where $\mathrm{Tr}\boldsymbol{R}_\Delta$ is the trace of $\boldsymbol{R}_\Delta$; the PSP (rotation axis) can be found as the eigenvector of $\boldsymbol{R}_\Delta$ that corresponds to an eigen value of 1 [26].

PSA Technique

The differences between the JME and PSA techniques occur only in the data analysis. Not only can the same experimental setup be used, but the same measurement procedure can be followed. The PSA technique works with derivatives of the measured Stokes vectors rather than the polarization transfer matrix. For linear polarization launch states at approximately 0°, 45°, and 90°, the corresponding output polarization states are measured at two closely spaced wavelengths. Vector products of these three states are used to form four output Stokes vectors **h**, **q**, **c**, and **c´**, such that the finite change due to wavelength of each, Δ**h**, Δ**q**, Δ**c**, and Δ**c´**, gives the DGD [27, 28]

$$\Delta\tau = \frac{1}{\Delta\omega}\arcsin\left[\frac{1}{2}\sqrt{\frac{1}{2}(\Delta\vec{h}^{2} + \Delta\vec{q}^{2} + \Delta\vec{c}^{2})}\right]$$

$$+\frac{1}{\Delta\omega}\arcsin\left[\frac{1}{2}\sqrt{\frac{1}{2}}(\Delta\vec{q}^{2}+\Delta\vec{v}^{2}+\Delta\vec{c}^{1\,2})\right] \quad (3.13)$$

The PSA also allows calculation of the PSP using the same four vectors. *PA, SOP, and PS:* Various other polarimetric measurement techniques with similarities to JME, PSA, and MMM are sometimes mentioned. But their names generally come up under the topic of "Other measurement techniques". These techniques, Poincaré Arc (PA), State of Polarization (SOP) and Poincaré Sphere (PS) are often mentioned without literature references and so, it is difficult to define a measurement procedure to associate with each name. A good generalization would be to use these somewhat generic titles to refer to techniques that measure the DGD by measuring $\Delta\mathbf{S}/\Delta\omega$ only. This requires the assumption that,

$$\left|\frac{dS}{d\omega}\right| = \left|\frac{d\theta}{d\omega}\right| \quad (3.14)$$

This is true when $\mathbf{S}(\omega)$ lies on a great circle, and occurs when an input polarization state is launched so that both PSP are equally illuminated. This condition is difficult to maintain in a mode coupled device such as a long fiber, but is possible in non-modecoupled components. As an example, a $\Delta\tau$ = 1 ps device would have an arc of ~ 45° for a wavelength change of 1 nm.

Fixed Analyzer

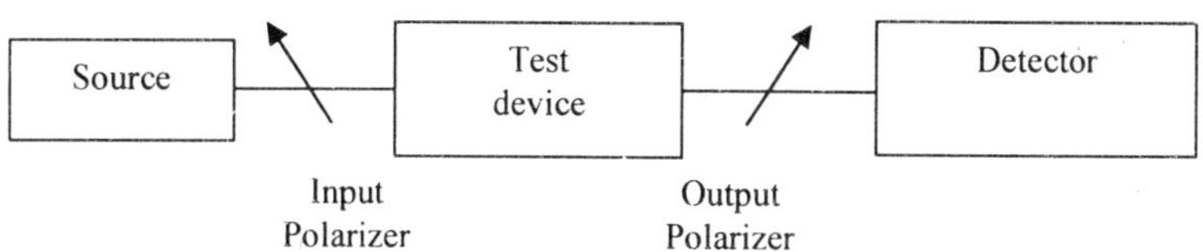

Fig. 3.6: Light Transmitted through Transmitter

A simple, but different approach for a polarimetric measurement technique in the frequency domain is the Fixed Analyzer (FA) method [29]. It is sometimes called Wavelength Scanning. The FA technique measures mean DGD based on Eq (3.10), but indirectly. Figure 3.6 illustrates the basic setup. Light transmitted through a polarizer - test device - polarizer setup is detected as a function of wavelength. This can be done either with a tunable laser and detector combination or with a broadband source and an optical spectrum analyzer (or monochromator). As the output polarization vector **S** moves around on the sphere, the intensity $I(\omega)$ transmitted through the output polarizer is proportional to the quantity $1+\sin(\theta)\sin(\varphi)\cos(\Phi)$, where Φ is the angle between Ω and Stokes vector describing the axis of the output polarizer, and φ is the angle between **S** and Ω (Fig. 3.5), and is independent of ω for non-mode-coupled devices. For non-mode-coupled devices, $\theta(\omega)$ depends approximately linearly on ω a nd contains all of the optical frequency dependence of $I(\omega)$, and so we can estimate $d\theta/d\omega$ by merely counting the number of extrema (peaks and valleys) in the sinusoidal $I(\omega)$ curve over a given optical frequency range. That is,

$$\langle \Delta\tau \rangle \omega_a - \omega_b = \frac{kN_e \Pi}{\omega_b - \omega_a} = \frac{kN_e \lambda_a \lambda_b}{2(\lambda_a - \lambda_b)c} \tag{3.15}$$

where the brackets indicate the average DGD measured over the frequency range from ω_a to ω_b, k is a mode-coupling constant, N_e is the number of peaks and valleys measured over the frequency range from ω_a to ω_b. The right-most expression in Equation is merely the same expression in terms of wavelength instead of optical frequency.

If we consider the case of a mode-coupled device, the situation becomes more complicated in that $d\theta/d\omega$

can depend strongly on ω, and Ω can also have a second-order dependence ω (Ω is still independent of ω to first order). This means that $I(\omega)$ no longer has a simple sinusoidal dependence but behaves in a quasi-random way. Fortunately, the mean DGD can still be estimated by counting peaks and valleys in the $I(\omega)$ spectrum. In this case, the coupling factor k becomes 0.805. k comes from a Monte Carlo simulation and allows an accurate estimate of mean DGD when the device is strongly mode-coupled [30].

RF Phase Shift

Another technique to be discussed resembles the time-of-flight measurement. Tunable laser intensity modulated at a frequency ranging from several tens of MHz to a few GHz. The modulated light passes through a polarization controller and then the test device and is then detected. The phase of the detected signal is compared to the phase of the modulator as the polarization state of the light is changed. The arrival phase φ_{RF} is related to the time of flight through the device, so that as the input polarization state is changed, the maximum and minimum phase measured $\varphi_{RF,max}$ and $\varphi_{RF,min}$ can be used to find the maximum and minimum polarization dependent propagation delays through the device. Their difference gives the Differential Group Delay through the test device,

$$\Delta\tau = \frac{\varphi_{RF,\max} - \varphi_{RF,\min}}{360^\circ . f_{\text{mod}}} \tag{3.16}$$

where phases are given in degrees and f_{mod} is the RF modulation frequency [31]. The down side of this intuitive approach is that it requires many phase measurements while the polarization state is randomly varied to find the maximum and minimum. This approach is improved by algorithms that allow $\Delta\tau$ to be obtained by measuring the RF phase at just four known polarization state. Two such implementations are the Modulation Phase Shift (MPS) technique [32] and Polarization dependent Signal Delay (PSD) method [33]. The PSD technique offers the advantage of calculating both the DGD and the PSP. RF phase shift is a promising technique in that it allows narrowband measurements whose spectral width is due to only the RF modulation frequency.

POSSIBLE SOLUTIONS ENABLING TO REDUCE POLARIZATION MODE DISPERSION

Four possible solutions will be considered for measuring Polarization Mode Dispersion in components: the Interferometric, Fixed Analyzer, and Stokes Parameter Evaluation techniques, which are usually used to measure PMD of single mode fibers, and the Phase Shift method, which is often associated with Chromatic Dispersion measurement of single mode fibers.

Interferometric Method

A basic measurement system for the Interferometric technique [34] is shown in Fig. 3.7 Light from a broadband source goes through a polarizer and then to the device under test. The output from the DUT is inputted to an interferometer a standard Michelson type is shown in here—which is then received by a detector.

The signal arriving at the detector is the autocorrelation of the electric field of the light, and in order to get good 'visibility', the polarizer must be aligned midway between the fast and slow axes of the DUT so that half of the light travels down each axis. If a component is made of a discrete birefringent element

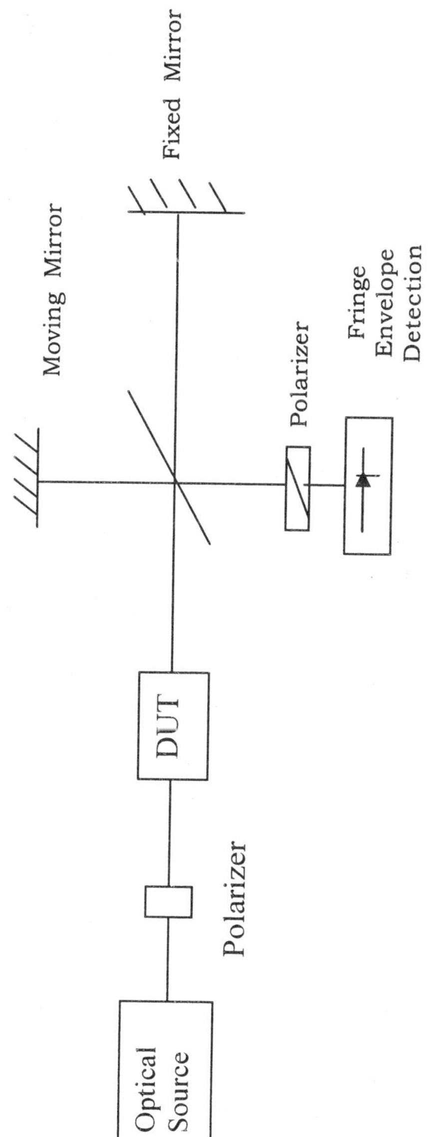

Figure 3.7: Interferometric PMD Measurement System with a Michelson Interferometer

then the device will resolve the light into two orthogonally polarized beams. As little or no mode coupling occurs, these polarization states will be maintained to the exit of the device, and light from the slow axis will arrive at a time $\Delta\tau$ later than from the fast. The analyzer axis, set midway between DUT's principal polarization states, passes light from each arm of the interferometer to the detector. In the interferometer, light from the fast axis will be spilt into two by the beam splitter and travel in the fixed and movable arms, as will the light from the slow axis. For zero optical path difference, interference will occur between light from the two arms of the interferometer. Interference will occur again between light from the fast and slow axes when light from the fast axis goes to the movable arm and is delayed by a time equivalent to the DGD, $\Delta\tau$, with the slow axis light in the fixed arm. Also, when light from the fast axis is in the fixed arm, the slow axis light must travel a shorter distance, hence shorter time compared with the fast axis light in the fixed arm, to interfere.

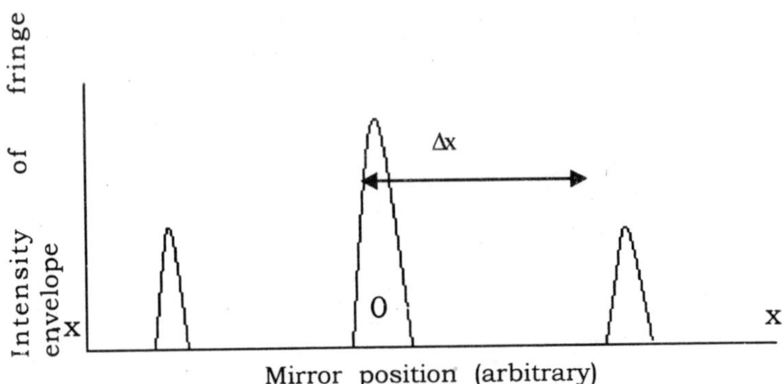

Fig. 3.8: Diagram to illustrate the intensity of the fringe envelop from an interferometer for a single element birefringent device

If the envelope of the intensity of the signal on the detector is plotted against distance, three peaks will be seen (Fig. 3.8). The central peak is at zero path difference, and the peaks either side of this are at a distance Δx away, which is equivalent to the DGD. Δτ is calculated using Eq (3.17).

$$\Delta\tau = 2\Delta x / c \tag{3.17}$$

Where c is the velocity of light in free space. The condition needed to measure PMD in this way is that the coherence time of the source is less than the PMD of the DUT [8], sourcetc < PMD. The source coherence time is given by Eq. (3.18)

$$_{source}t_c = \lambda_c^2 / \Delta\lambda_c \tag{3.18}$$

where λ_c is the central wavelength of the source, Δλ is the source spectral width, Full Width Half Maximum. When there is more than one birefringent element in a device, additional peaks will be present on the plot. If any of these peaks are non-PMD related they must be excluded from calculations. They can be identified as those that do not change in amplitude as the input lead configuration or input polarization state is changed.

Fixed Analyzer

The Fixed Analyzer, or wavelength scanning method measures PMD by scanning through different wavelengths, and looking at the intensity of the signal on a detector after it has passed through a fixed polarizer, the device under test, and a further fixed polarizer, the analyzer [35, 36]. This setup is shown in Fig. 3.9

For each wavelength, light inputted to the fiber will be changed to an unknown polarization state after

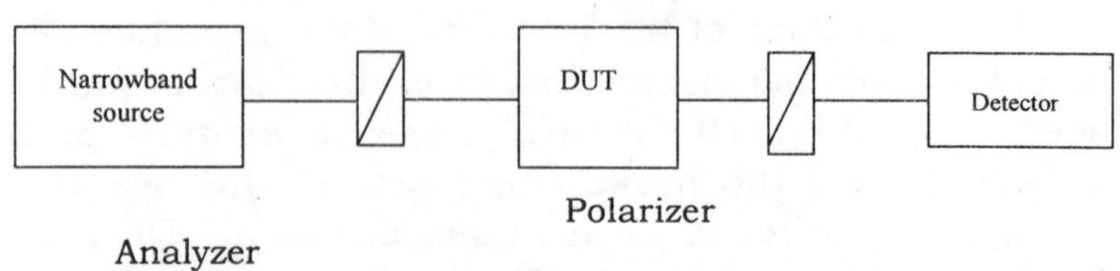

Fig. 3.9: PMD Measurement System for the Fixed Analyzer Technique

having traveled down its length. At the fixed analyzer only a fraction of the output light will pass through, according to the offset of this state with the analyzer axis. A reference measurement is made with either the device removed or the analyzer rotated by 90 degrees.

A ratio is then taken of the intensity of light with the analyzer in place over the reference value, and plotted against wavelength. Peaks and troughs (extreme) are seen, which, for a discrete birefringent component, follow a sinusoidal pattern. Determining the PMD involves either counting the number of extrema, Ne, for a given wavelength interval, or alternatively, carrying out a Fourier Transform of the intensity plot [37, 38]. Analysis using extrema counting (EC) or Fourier Transform gives the same result, and so the formulae in this report relate to extrema counting as they are more transparent. PMD in components is found from Eq. (3.19) with the low mode coupling regime, k = 1.

$$\langle \Delta\tau \rangle = \frac{KE\lambda_1\lambda_2}{2(\lambda_2\lambda_1)c} \tag{3.19}$$

where $\langle \Delta\tau \rangle$ PMD

k mode coupling factor, k = 1 for no mode coupling, 0.82 for strong mode coupling,

λ_1 lower wavelength,

λ_2 upper wavelength,

E E = Ne, Number of extrema, for the wavelength range λ_1 to λ_2, E = (Ne −1) when λ_1 and λ_2 coincide with extrema,

C speed of light in vacuum.

Equation (3.19) can alternatively be expressed in terms of optical frequency:

$$\langle \Delta\tau \rangle = \frac{kE\Pi}{\Delta\omega} \quad ....3.20$$

where $\Delta\omega$ optical frequency window of measurement in radians per second.

In exploring the capabilities of this technique, the number of extrema was set to 2, and in order to facilitate estimating the PMD for different wavelength ranges, equation (3.19) was approximated to (3.21). For all of the calculations in, the central wavelength, λ_c, was kept constant, at 1550 nm.

$$\langle \Delta\tau \rangle = \frac{kE\lambda_c^2}{2(\lambda_2 - \lambda_1)c} \quad (3.21)$$

It is apparent that lowering the PMD measured would mean an increase in the wavelength range scanned, as the extrema are further apart. As the PMD here is deterministic, the spacing between the peaks and troughs is fixed. A sufficient range must therefore be scanned to ensure that at least two extrema have been encompassed, which for a PMD of 0.1 ps means scanning over a minimum of 80 nm. This makes it impractical if the available bandwidth of the device is only 0.2 nm. The wavelength scanning method is limited to detecting a minimum PMD of 40 ps for a DUT with extrema separated by 0.2 nm, giving this technique a figure of merit.

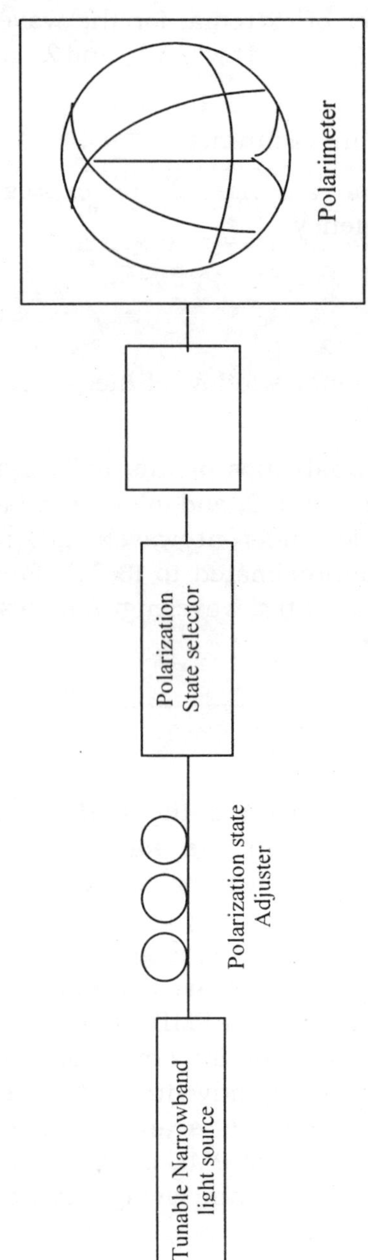

Figure 3.10: PMD Measurement System for Stokes Parameter Evaluation

The fundamental limitations of the technique are the same whether analysis is carried out by extrema Counting or Fourier Transform. However, a benefit of Fourier Transform is that a plot is obtained from this process, much like that from the Interferometric technique, which displays multiple peaks if more than one birefringent element is within the device. It distinguishes different sources of PMD.

Stokes Parameter Evaluation

Stokes Parameter Evaluation is a generic term for the frequency domain measurement techniques of Jones Matrix eigen analysis (JME), State of Polarization (SOP) and Poincaré Sphere Analysis (PSA). The basic principle common to all of these is that the Stokes parameters are used to directly determine the Differential Group Delay. The JME and PSA methods are equivalent, and the SOP is a simplified version of PSA.

Figure 3.10 shows the experimental layout, which is the same for all three methods. A tunable narrowband source, such as a laser, is used to provide light that remains polarized for all conditions. The polarization state is defined using a polarization state selector. For the JME and PSA methods, three linear states that are mutually orthogonal on the Poincaré sphere are launched in turn into the DUT at each wavelength point scanned, and for the SOP, one state is launched. A polarimeter then determines the Stokes Parameters of the light exiting the device as the polarization state of the light has been altered from its input state by the transmission properties of the fiber.

An output polarization state is represented by a point on the Poincaré sphere. As the wavelength changes for a given input polarization, the output polarization state changes and moves to a different point on the sphere.

For a discrete birefringent component, a circle is swept out on the sphere as the wavelength is incremented. The rate of change of polarization state with optical frequency is an indication of the level of PMD. Small PMD's will cause a slower precession around the PSP axis than larger PMD's for an equivalent change in optical frequency. A separate circle is drawn out for each input polarization state launched.

Analysis can be performed by Jones Matrix eigen analysis [39], in which the DGD is calculated from equation (9), and involves formulating a transmission Jones matrix from the Jones vectors of two adjacent wavelength points. This is only valid if there is no PDL present. The DGD is quoted at the midpoint of the wavelengths.

$$\Delta\tau = \left| \frac{Arg\left(\frac{\rho_1}{\rho_2}\right)}{\Delta\omega} \right| \qquad (3.22)$$

where:

$\Delta\tau$	DGD,
ρ_1, ρ_2	complex eigen values of $\mathbf{T}(\omega + \Delta\omega)\,\mathbf{T}^{-1}(\Delta)$,
$\mathbf{T}(\omega + \Delta\omega)$	Transmission Jones matrix, at the higher optical frequency,
$\mathbf{T}^{-1}(\omega)$	Inverse Jones matrix, at the lower optical frequency,
Arg	denoted the argument function, that is $Arg(\eta e^{i\theta}) = \theta$.
$\Delta\ \omega$	optical frequency step, expressed in radians per second.

With Poincaré Sphere Analysis [40], DGD is also calculated using matrix algebra (3.23).

$$\Delta\tau = \frac{1}{\Delta\omega}\left[\arcsin\left(\frac{1}{2}\sqrt{\frac{1}{2}}(\Delta\vec{h}^{2} + \Delta\vec{q}^{2} + \Delta\vec{c}^{2})\right)\right.$$

$$\left. + \arcsin\left[\frac{1}{2}\sqrt{\frac{1}{2}}\left((\Delta\vec{q}^{2} + \Delta\vec{v}^{2} + \Delta\vec{c^{1}}^{2})\right)\right]\right] \qquad (3.23)$$

where:

Δ denotes a difference between two adjacent wavelength points,

h, v, q, Stokes vectors, independent of input SOP,

c and c^{l} and vector products formed from the Stokes vectors.

For the State of Polarization method, DGD is found from the phase difference of light exiting the DUT as a result of a frequency change of the optical source, and is given by the angle subtended on the Poincaré sphere. As SOP has the most transparent calculations of the three techniques its equations are used for numerical analysis in this report. It is, however, less robust than JME and PSA, and only works well when the change in output polarization state is not rapid and random, which poses no problems for application to components. Equation (3.24) defines the DGD in terms of radians.

$$\Delta\tau = \frac{\Delta\theta}{\Delta\omega} \qquad (3.24)$$

where:

$\Delta\tau$ differential group delay in seconds

$\Delta\theta$ rotation about the principal states axis in radians

$\Delta\omega$ radian optical frequency change

This is rewritten in terms of optical frequency and wavelength in Eq. (3.25].

$$\Delta\tau = \frac{\Delta\theta\lambda_1\lambda_n}{2\Pi c\Delta\lambda\Pi} = \frac{\Delta\theta}{2\Pi\ \Delta f} \quad (3.25)$$

where:

λ_1 wavelength at frequency f_1,

λ_n wavelength at frequency f_n,

Δf change in frequency,

c speed of light in a vacuum.

When the DGD for each wavelength pair has been calculated, a histogram is then plotted of the frequency of occurrence of each DGD value. For a component with discrete birefringence, this should be single valued. The delay corresponding to the peak value of a curve fitted to the distribution is defined as the PMD of the device.

CONCLUSION

In the four possible solution obtained for measuring PMD in the following technique the interferometric method, the fixed analyzer, and stokes parameter technique which are usually to measure PMD of the single mode fiber the interferometric technique has been chosen to be the best technique on the basis of the availability of the resources and it is noted that the propagation delay time due to PMD in this technique is less while comparing to other techniques shown in the chapter. The details of the interferometric technique has been discussed in Chapter 4.

REFERENCES

1. Rashleigh, S. C. & Ulrich, M. (1978) *Opt. Lett.* **3**, 60–62.
2. Poole, C. D. & Wagner, R. E. (1986) *Elect. Lett.* **22**, 1029–1030.
3. Poole, C. D. & Nagel, J. A. (1997) in *Optical Fiber Telecommunications*, eds. Kaminow, I. P. & Koch, T. L. (Academic, San Diego), Vol. IIIA, pp. 114–161.
4. Heismann, F. (1998) *ECOC'98 Digest* **2** (Sept.), 51–79.
5. Menyuk, C. R. (1999) *J. Eng. Math.* **36**, 113–136.
6. Hecht, E. Optics, USA, Addison-Wesley, 1987 (2nd edition).
7. Derickson, D., Editor. Fiber Optic Test and Measurement, Upper Saddle River, New Jersey 07458, Prentice-Hall PTR 1998.
8. Poole, C.D. and WAGNER, R.E. Phenomenological approach to polarisation dispersion in long single-mode fibers. *Electron. Lett.*, 1986, **22**, 19, 1029 - 1030.
9. Huttner, B and GISIN, N. Anomalous pulse spreading in birefringent optical fibers with polarization-dependent losses. *Opt. Lett.*, 1997, **22**, 8, 504 - 506.
10. Gisin, N. and Huttner, B. Combined effects of polarization mode dispersion and polarization dependent losses in optical fibers. *Opt. Commun.*, 1997, **142**, 119 - 125.
11. CCITT Recommendation G.651 - Characteristics of a 50/126 micron graded index optical fibre cable. *COM. XV-R-89 E,* May 1992.
12. N. Gissin, R. Passy, J. C. Boschoff and B. Perny, *IEEE Photon. Tech. Lett.*, vol. 5, p. 819, 1993.
13. N. Gisin, "Polarization Mode Dispersion: Definitions, Measurements and Statistics", OFMC' 93, Torino, pp. 149 - 154, 1993
14. C.D. Pole *et al., Optics Letters,* vol. 13, pp. 155 - 157, 1988
15. F. Corsi, A. Galtarosa and L. Palmieri, "Polarization Mode Dispersion Characterization of Single-Mode Optical Fibers Using Backscattering Technique", *Journal of Light wave technol.*, vol. 16, no. 10, pp. 1832 - 1843, 1998

16. C.D. Poole, and C.R. Giles, "Polarization-dependent pulse compression and broadening due to polarization dispersion indispersion-shifted fiber", *Opt. Lett.*, **13**,155-157 (1987).

17. Yoshinori Namihira and Jun Maeda,"Polarization mode dispersion measurements in optical fibers", *Tech. Dig. – Symp.on Opt. Fiber Meas.*, Boulder, 145-150 (1992).

18. Bamdad Bakhshi, et al., "Measurementof the Differential Group Delay in Installed Optical Fibers Using Polarization Multiplexed Solitons," *IEEE Phot.Tech. Lett.*, **11**, 593-595 (1999).

19. Joseph W. Goodman, *Statistical Optics*,(Wiley, New York, 1985) p. 168.

20. B.L Heffner, "Influence of optical source characteristics on the measurement of polarization-mode dispersion of highly mode-coupled fibers", *Opt. Lett.*, **21**, 113- 115, (1996).

21. TIA/EIA FOTP-124. 1999. *Polarization- mode dispersion measurement for single- mode optical fibers by interferometric method.* Telecommunications Industry Association, Arlington, VA.

22. C.D. Poole et al., "Polarization dispersion and principal states in a 147 km undersea lightwave cable," *J. of Lightwave Tech.*, **LT-7**, 1185-1190 (1989).

23. B.L. Heffner, "Automated Measurement of Polarization Mode Dispersion Using Jones Matrix Eigenanalysis," *IEEE Photonics Tech. Lett.*, **4**, 1066-1069 (1992).

24. R. Clark Jones, "A New Calculus for the Treatment of Optical Systems: VI. Experimental Determination of the Matrix," *J.O.S.A.* **37**, 110-112 (1946).

25. R.M. Jopson, "Measurement of Second- Order Polarization-Mode Dispersion Vectors in Optical Fibers," *IEEE Photonics Tech. Lett.* **11**, 1153-1155 (1999).

26. Mary L. Boas, *Mathematical Methods in the Physical Sciences* (Wiley, New York, 1983) p.454.

27. Normand Cyr, *et al.*, "Stokes Parameter Analysis Method, the Consolidated Test Method for PMD Measurements," *Proceedings Nat. Fib. Opt. Eng. Conf.*, Chicago, (1999).

28. TIA/EIA FOTP-122. 1999. *Polarization-Mode Dispersion Measurement for Single-ModeOptical Fibers by Stokes Parameter*

Evaluation. Telecommunications Industry Association, Arlington, VA.

29. Craig D. Poole, "Polarization-Mode Dispersion Measurements Based on Transmission Spectra Through a Polarizer," *J. Lightwave Tech.*, **12**, 917-929 (1994).

30. P.A. Williams and C.M. Wang, "Corrections to Fixed Analyzer measurements of Polarization Mode Dispersion," *Journal of Lightwave Technology,* **16**, 534-541 (1998).

31. P.A. Williams *et al.*, "Narrowband measurements of polarization-mode dispersion using the modulation phase shift technique," *Tech. Dig. - Symp. on Opt. Fib. Meas.*, Boulder, 23-26 (1998).

32. P.A. Williams, "Modulation phaseshift measurement of PMD using only four launched polarization states: a new algorithm," *Elec. Lett.*, **35**, 1578-1579 1999).

33. Lynn E. Nelson, et. al. "Measurement of polarization mode dispersion vectors using the polarization-dependent signal delay method," *Opt. Express,* **6**, 158-167 (2000).

34. American Standard, TIA FOTP-124:1999. Polarization-Mode Dispersion Measurement for Single-Mode Optical Fibers by Interferometry.

35. Gisin, N. and Pellaux, J.P. Polarization mode dispersion: time versus frequency domains. *Opt. Commun.*, 1992, **89**, 316-323.

36. Poole, C.D. and Favin, D.L. Polarization-Mode Dispersion Measurements Based on Transmission Spectra Through a Polarizer. *J. Lightwave Tech.*, 1994, **12**, 6, 917 - 929.

37. American Standard, TIA FOTP-113:1996. Polarization-Mode Dispersion Measurement for Single-Mode Optical Fibers by the Fixed Analyzer Method.

38. IEC 60793-1-48:to be published. Polarisation mode dispersion measurement methods.

39. IEC 61300-3-32:to be published. Polarisation mode dispersion for passive optical components.

40. American Standard, TIA FOTP-122:1996. Polarization-Mode Dispersion Measurement for Single-Mode Optical Fibers by Jones Matrix Eigenanalysis.

4

Introferometric Technique to Reduce PMD in Optical Fiber

INTRODUCTION

Single mode fiber supports two orthogonal polarization modes. In a perfectly circular core fiber, each of these modes experiences the same propagation delay. In real fiber, deformations from the ideal shape occur, resulting in a difference in the propagation speed of each of the two polarization modes. The associated difference in the propagation delays of these modes broadens any pulse sent through the fiber and impairs transmission. The difference in propagation delays is known as polarization mode dispersion. The extent of PMD within a fiber varies from manufacturer to manufacturer. It arises from fiber core deformations. Such deformations are inevitably produced during the fiber manufacture, but are also the result of stress on a fiber when it is packaged into a cable, and manipulated during cable laying operations. Polarization Mode Dispersion is mainly characterized by the differential group delay (DGD) between the fast and slow states of polarization (SOP). If this delay difference is a significant fraction of a symbol period, then PMD can be expected to seriously degrade the performance of the transmission link over what it would be without PMD. For the non-return to zero (NRZ) binary data transmission schemes typically utilized in optical links, we can regard PMD to be a significant problem when the DGD approaches one-tenth of a bit duration, as a rule of thumb. In many optical fiber, the mean DGD does not exceed 40 ps, which is one tenth of an OC-48 bit-period. However, OC-192 transmission, with a bit duration of 100 ps, is jeopardized for fibers having a mean DGD of 40 ps or more. While the DGD dong a fiber is a fixed function at any instant, due to the random origins of the fiber geometric disturbances, over the ensemble of like fibers the DGD may be viewed as a random process evolving over time, wavelength, and various their parameters. With respect to time and wavelength, we can see that the DGD can be expected to be stationary (and thus would have a mean which does not Vary with these parameters). The DGD has a Maxwellian distribution, which is uniquely characterized by its mean. The random nature of PMD also requires that the variation of DGD over time and wavelength be characterized, although the literature has neglected this aspect so far.

The PMD of a system consisting of a cascade of optical fibre components is a function of the PMD of each of its components. If we assume the statistics of the PMD of each subsystem to be statistically unrelated, then we can derive an expression for the PMD of the total system in terms of the PMD of the subcomponents. The validity of the assumption and thus the expression may be questioned. Several methods of measuring PMD exist. The methods vary in their practicality and utility. Since the goal of this work cannot be attained without measurements, a discussion of the major PMD measurement techniques is given in Chapter 4. From these possible techniques, a method is selected and then used for measurements of PMD reported in the subsequent sections of the work The general structure of an optical fiber is a core with a high index of refraction, surrounded by a lower index cladding. Transport fiber is made of fused silica with a refractive index of approximately 1.5. When a wave propagates along the fiber core and approaches the interface between the two mediums, the lower refractive index permits total internal reflection of the wave back towards the core. This is the basic principle of how a fiber serves as an optical waveguide. An optical mode is a particular solution of the electro-magnetic wave equation which satisfies the boundary conditions pertinent to a propagation medium ([1], p. 32). Optical fiber modes can generally either be guided, such that they propagate with relatively little attenuation, or radiative, such that light is absorbed into the cladding.

Multi-mode fiber has a wide core such that the wave equation has many guided modes. An applied signal will produce components in many or all of these modes. The different modes propagate along different spatially distributed paths, making full use of total internal reflection as shown in Fig. 4.1(a). Since any two such paths will generally have different lengths, the wave-fronts of components dong these two modes will arrive out of phase at the exit. The two modes effectively propagate at different speeds along the length of the fiber, and the disparity in arrival times of a signal is defined as multi-mode dispersion.

As the name suggests, single mode fiber (SMF) was developed to support only a single guided mode along which a light wave can travel successfully. The core of the SMF is made sufficiently small that the only solutions of the wave equation are (a) a guided mode and (b) multiple leaky modes. At the exit of the fiber, only the guided mode persists, as illustrated in Fig. 4.1(b) (the curved lines represent wave-fronts). In the expressions "multi-mode fiber" and "single mode fiber", the modes referred to are named "spatial modes" since the waves propagating along them can be thought of as taking different paths. The elimination of multi-mode dispersion through the use of SMF is key to high bit-rate digital communications.

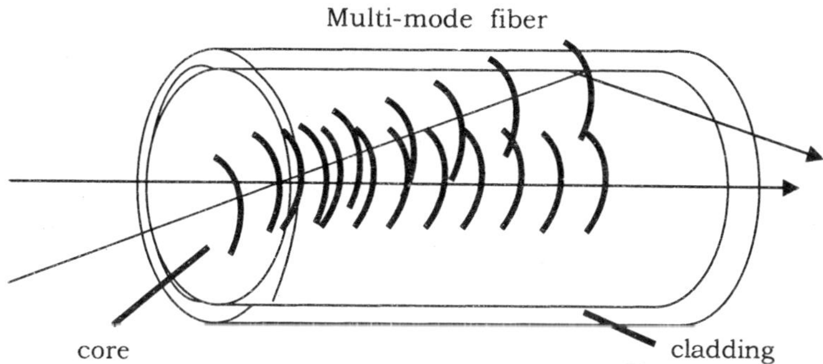

Fig. 4.1(a): Single-mode Fibre

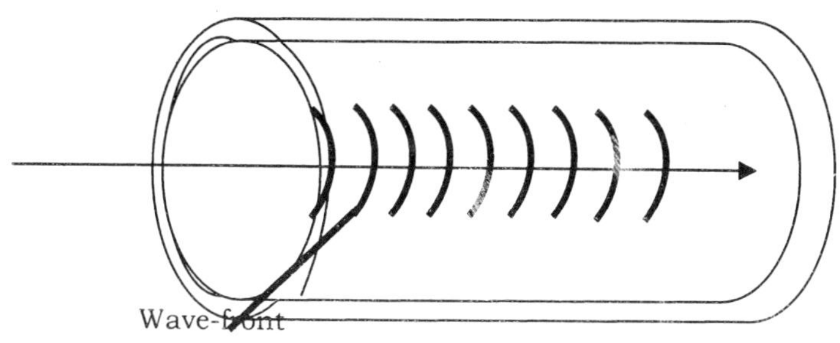

Fig. 4.1(b): Muiti-Mode vs. Single Mode Fiber

However, in a real fiber, deformations of the core geometry exist. Ellipticity in the fiber core causes birefringence, a difference in refractive index from one polarization mode to the other. Since the speed of light is inversely proportional to the index of refraction, a light wave containing components in both polarization modes launched at one end of the fiber will propagate with two different speeds, and the modes will arrive out of phase at the far end. This phase difference is the result of a propagation delay difference which is the main effect of polarization mode dispersion. naturally, this effect will cause signal degradation in a digital communications system. Figure 4.2 shows the physical model of PMD within a single mode fiber. The birefringence in each section is characterized by the differential time delay between "$\Delta\tau$ both polarization modes.

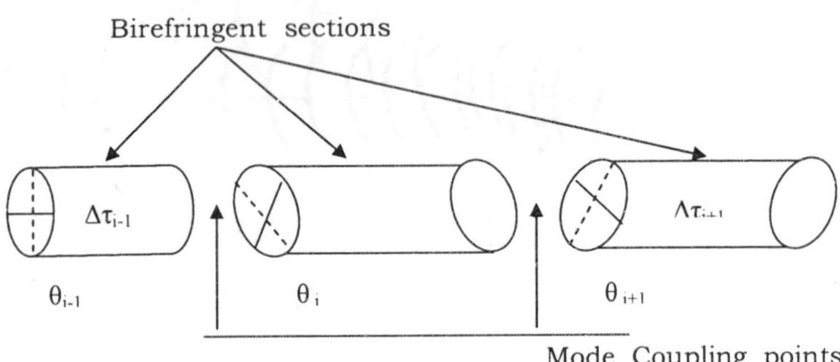

Fig.4.2: SMF Physical Model

In this model, the axes in one section are randomly oriented with respect to the axes in the next section; in other words, the orientations from section to section are uncorrelated. From this assumption of uncorrelatedness between sections, the delays add up as the root sum of squares (RSS), i.e., the total differential group delay in a fiber is the square-root of the sum of the squares of the delays in all of its birefringent sections.

The above model is in fact a simplification of the birefringent structure in a real fiber. In uncorrelated sections within a fiber, the core geometry is generally not uniform over the length of each section, and as a result, the sections do not have uniform birefringence. Each section would then be described by incident and exit mode coupling angles.

MEASURING PMD

Several methods have been devised to measure the PMD within an optical fiber component. These methods vary in complexity and utility. Three methods will be discussed here, namely the Jones matrix eigen-analysis method, the wavelength scanning method and the interferometric method. The theory behind them, the equipment setup, dynamic range and step size will be discussed, and typical results from their application given.. Based on the latter fluctuation, an expression is given to test the real-world validity of a DGD measurement, knowing the bandwidth over which the measurement is taken. The goodness of fit test used on the measured distributions is also explained.

Interferometric Method

Interferometry is a method of measuring the autocorrelation of an optical electrical field with respect to time, using the interference pattern created when adding a waveform to a delayed copy of itself. The interferometer splits an incoming light wave along two paths, and recombines the light at the output. The optical intensity of the recombined wave is measured while varying the spatial length of one path with respect to the other. The differential spatial length results in a time delay between the two paths, and is controlled through the movement of a mirror [2]. For a perfect monochromatic wave, the interferogram over delay resembles the coherence spectral characteristics of the source over time centered about O. The autocorrelation peak appears when the spatial delay between the two paths is null. However, such a monochromatic wave is highly coherent, and will therefore have a large coherence time, which is reflected in the FWHM of the autocorrelation peak. A spectrally wide source with a small coherence time will have a narrow autocorrelation peak. As mentioned earlier, PMD causes an optical wave to be split into two polarization states, adds a differential delay between the states, and recombines the two waves at the output. Therefore, at the exit of a fiber, if a polarization mode can be recombined with its orthogonal mode, and assuming the polarization modes are constant over the spectral bandwidth, the inter-ferogram of a spectrally wide optical wave propagating through the fiber is expected to have, in addition to a central autocorrelation peak at a spatial delay of 0, two autocorrelation peaks where the magnitude of the spatial delay is equivalent to the value of the DGD. This is the principle behind the interferometric PMD measurement technique. If a source with a wide spectral width, such as a light-emitting diode (LED),is used, peaks appear at spatial delays equivalent to al1 of the DGDs within the spectral range of the source. In the case of a polarization maintaining fiber, the DGD is constant over wavelength, and thus a perfect PMD interferogram for this device has two predominant peaks besides the central autocorrelation peak. More autocorrelation peaks show up in the presence of random mode coupling.

Experimental Setup

The interferogram is based on the phenomenon of irradiance. The intensity of the fringes of an inter-ferograrn are proportional to the dot product of the resultant electric field vector by its complex conjugate. At a given location in space, if an electrical wave has components along two orthogonal States of polarization, the dot product of one component of the field by the complex conjugate of the other is null due to the orthogonality of the vectors. To combine the electrical field in one polarization state with the field in the orthogonal mode, a method must be put into place to rotate the polarization of one path of the interferometer with respect to the other. In the setup shown in Figure 4.3, a 1/8th wave plate is used. A wave plate is an optical device made of birefringent material which has two linear axes perpendicular to the direction of propagation, namely the fast and slow axes. When *an* optical wave propagates through the plate, the component of the wave along the slow axis is retarded with respect to the component dong the fast axis, by a phase dictated by the thickness of the wave-plate. The Jones matrix of a wave plate is:

$$T_{\text{wave plate}} = \begin{pmatrix} 1 & 0 \\ 0 & \exp(-j\Gamma) \end{pmatrix} \tag{4.1}$$

where Γ is the phase (in radians) by which the slow axis is retarded over the fast axis $\Gamma = 2\Pi/8 = \Pi/4$).

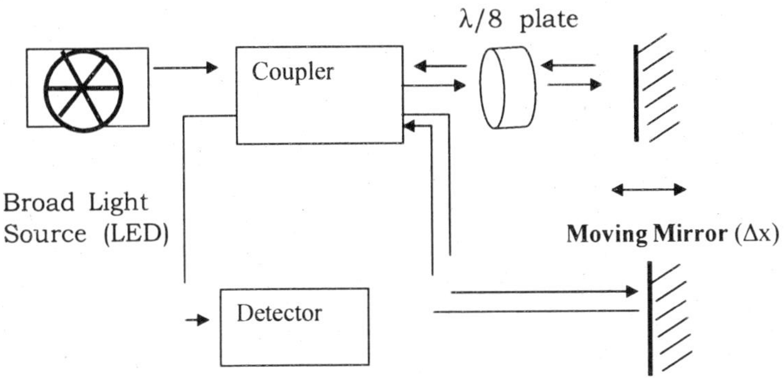

Fig. 4.3 Interferometer Measurement Setup

When used in conjunction with a mirror, whose Jones matrix is, ideally:

$$T_{\text{mirror}} = \begin{pmatrix} -1 & 0 \\ 0 & 1 \end{pmatrix} \tag{4.2}$$

the 1/8th wave plate acts as a quarter-wave plate: a light wave propagating through the 1/8th wave plate, reflected off the mirror, and propagating through the same wave plate in the opposite direction, as described in Fig. 4.3, will have effectively passed through the equivalent of a quarter-wave plate:

$$Tequiv = T_{\lambda/8}.Tmirror.T_{\lambda/8} = \begin{pmatrix} 1 & 0 \\ 0 & e^{j\frac{\Pi}{4}} \end{pmatrix} \cdot \begin{pmatrix} -1 & 0 \\ 0 & 1 \end{pmatrix} \cdot \begin{pmatrix} 1 & 0 \\ 0 & e^{-j\frac{\Pi}{4}} \end{pmatrix}$$

$$= -1\begin{pmatrix} 1 & 0 \\ 0 & j \end{pmatrix} \tag{4.3}$$

while the optical wave travelling through the other arm of the interferometer will be subject to the polarization transformation. The experimental setup shown in Figure 4.3 resembles the one used in [3].

An LED is used as the source at the input of the device under test. Linearly polarized light, making an angle of 45^0 with respect to the two axes of the plate, is transformed by the effective quarter-wave plate into circularly polarized light (the converse applies for incident circularly polarized light). Since circularly polarized light can be broken down into linearly polarized components (and vice-versa), if the output PSPs of the fiber have linear (at 45^0) or circular polarization States, then, as the two arms of the interferometer recombine, the delayed signal component will have a polarization component parallel to the non-delayed signal component. Therefore, at a fixed wavelength, as long as both output PSPs of the fiber each have components along the two axes of the interferometer's wave plate, and as long as the light source has components along the two PSPs, the optical source component along the fast PSP can be partially recombined with the delayed component along the other PSP, and the DGD at that wavelength will correspond to a set of peaks in the interferogram. Fortunately, in highly mode-coupled fibers, the wide spectrum of the LED will explore a wide range of PSPs, such that the interferogram will contain several peaks corresponding to various DGDs. This reduces the necessity to monitor launch conditions of the optical source into the fiber.

The envelope of the resulting interferogram resembles a Gaussian. The half-width of this Gaussian can be shown to be related to the mean of the square DGD [2]. As mentioned, if the source is spectrally wide, then a multitude of DGDs will be explored by the interferometer over the frequency range of the source. The resulting interferograrn will have the expected central peak superimposed upon a Gaussian representing the various DGDs [2]. The mean DGD of a highly mode coupled fiber is measured by calculating the second moment of the Gaussian after removal of the central peak. Due to the limited spectrum of the LED, the uncertainty in the estimate of the mean DGD is identical to that calculated using other methods. It is only necessary to substitute the spectral width of the source for the wavelength range used with the JME or wavelength scanning methods. In addition, interferometric measurements contain additive noise and random amplitude as explained in [3], which must be compensated for. The central autocorrelation peak must also be eliminated.

In the setup described above, assuming nominally equal branch distances in each arm of the interferometer, such that the waves propagating down the two branches are subject to approximately the same attenuation, the mirror displacement Δx is related to the delay τ by

$$\tau = 2\frac{\Delta x}{c} \tag{4.4}$$

Therefore, in order to measure DGDs above ô ps, the mirror displacement must be capable of exceeding where **c** is the speed of light in air.

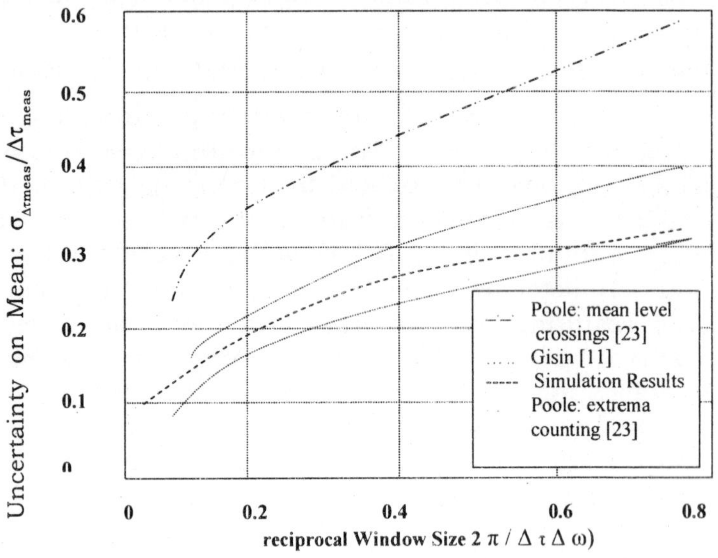

Figure 4.4 Uncertainty of the Estimate of the Mean

The equipment needed to perform an interferometric measurement was not available for the measurements performed in this thesis. The interferogram has an analogous equivalent to the Fourier transform of the transmission spectra of the wavelength scanning method. The main difference is the absence of the central peak corresponding to the autocorrelation of the unperturbed source. The LED source used with the interferometric method has a wide spectral width, while the source used with the wavelength scanning technique is spectrally narrow, and will have an approximately constant autocorrelation value over the time scale of interest.

RESULT AND DISCUSSION

In measuring PMD, the first choice to be made is deciding which measurement technique to use. Here, the various techniques have been described so as to illustrate their relative merits. Generally, decisions are made based on measurement time, spectral resolution, and what quantities are measurable. Once a measurement technique is chosen, it is most important to understand the sources of measurement error (and minimize them).

Four possible solutions will be considered for measuring Polarization Mode Dispersion in components: the Interferometric, Fixed Analyzer, and Stokes Parameter Evaluation techniques, which are usually used to measure PMD of single mode fiber, and the Phase Shift method, which is often associated with Chromatic Dispersion measurement of single mode fibers.

The basic measurement system of PMD is the interferometric technique the interferometric technique is capable of measuring PMD of components. If a component has an operational bandwidth of just 0.2 nm then the coherence time of the source is such that the minimum PMD measurable is 40 ps. If the coherence source having wavelength of 10 nm then the PMD is reduced to 0.08 ps. Further it is concluded that as a working coherence wavelength of a source is increased then the PMD is reduced.

Fixed analyzer is another technique to provide the solution to PMD. In exploring the capacity of this technique the PMD for a different wavelength ranges from 80 nm, 8 nm, 0.2 nm and 0.08 nm are used in calculating the PMDs. The wavelength 80 nm, 8 nm, 0.2 nm and 0.08 nm corresponds to PMD values 0.1ps, 1ps, 40ps, and 100ps respectively. With this it is concluded that as the wavelength decreases then the PMD value increases.

Again the above shown conclusion is conformed by stokes parameter evaluation. Stokes parameter evaluation is the general term for the frequency domain measurement technique of JME value analysis the basic principle of stokes parameters are used to determine the differential group delay and can be use to provide a definite answer, when there are complicity arises in the fixed analyzer and interferometric technique. It is derive that the wavelength 0.2 nm correspondence to the PMD 191 fs.

REFERENCES

1. G.P. Agrawal, Fiber-Optic Communications Systems, New York: John Wiley & Sons, 1997.
2. N. Gisin, J. P. Von der Weid, J. P. Pellaux, "Polarisation mode dispersion of short and long single-mode fibers," Journal of Light wave Technology, Vol. 9, No. 7, pp. 82 1-827, JUIY 199 1.
3. B.L. Heffner, "Compensation formula for noise threshold bias of interferometric PMD measurement," in NIST Technical Digest: Symposium on Optical Fiber Measurements 1996, Vol. 905, pp. 135-138, 1996.

5

Conclusions and Suggestions

The PMD is the limiting factor for the high bit rate optical transmission systems. The precise measurement of the PMD is important for the design and maintenance of these systems. They are several methods developed to this time for the PMD measurements. They differ mutually in many aspects - precision, reliability, repeatability, stability, prize and the possibility of use in field conditions. It allows not only to measure precisely the PMD but it also gives the complete view on the distribution of the total birefringence along the fiber.

Apart from the several techniques shown in our work the interferometric technique is capable of measuring PMD of components. However, it relies on using a broadband source without ripples, which severely limits the techniques effectiveness in characterizing PMD in components of restricted wavelength and it is concluded that if a component as a operational bandwidth of 0.2 nm then the PMD measured is 40 ps. If the coherence source having working wavelength of 100 nm then the PMD reduced to 0.08 ps. Further it is confirmed by the method of interferometric that as a working coherence source is increased then the PMD is reduced by means of which the speed of the communication of the optical signal through the optical fiber can be increased.

. SUGGESTIONS FOR FUPTHER WORK

Using the same order in which topics were discussed in this work, the following future work suggestions are made.

From the modelling point of view, an analytical derivation of the correlation bandwidth would give confidence to the empirical result. Along with that, analytical derivations of the simulation results pertaining to variance on the mean and cross-correlation error would be desirable. The measurements conducted in this thesis are laborious. For a Iong term PMD characterization test-bed, the processes of (a) determining the correlation temperature of a device and (b) generating independent PMD instances for a device and measuring The mean DGD should be automated, now that a methodology is in place. This would help narrow down the uncertainty on a device's mean DGD. In terms of statistical model characterization, automated measurements would permit a population of DGD samples large enough to characterize the tail of the distribution to be more easily obtained over a given period of time. This would help ascertain to a further degree the theoretical Maxwellian distribution. It would appear that the possibilities of numerical simulations are endless. From this perspective, a pulse of given duration could be made to propagate through the fiber already created by the simulation program. After addition of this feature to the program, the pulse distorting effects could be examined, and consequential communications penalty derived and compared to that of the literature. A study of PMD compensation could ensue from this. Finally, given the limitations of the devices which were measured, and the knowledge that the major DGD contributors were devices which are usually passive, a test of a system with high PMD fiber spans would be ideal. In this case, the DCMs could be held at constant temperature, and only the PMD in fiber spools would have to vary over time (or temperature). Due to the high DGD, fewer measurements would have to be taken to obtain mean DGD estimates with the same confidence interval as obtained in this thesis. From the systems perspective, a link with high DGD could be used to further study the effects of first and second-order PMD over time.

Bibliography

Agrawal, G.P., Fiber-Optic Communications Systems, New York: John Wiley & Sons, 1997.

American Standard, TIA FOTP-113:1996. Polarization-Mode Dispersion Measurement for Single-Mode Optical Fibers by the Fixed Analyzer Method.

American Standard, TIA FOTP-122:1996. Polarization-Mode Dispersion Measurement for Single-Mode Optical Fibers by Jones Matrix Eigenanalysis.

American Standard, TIA FOTP-124:1999. Polarization-Mode Dispersion Measurement for Single-Mode Optical Fibers by Interferometry.

Anderson, D., *Proc. Inst.Electr.Engg.* Vol. 123, 1985

Bakhshi, Bamdad et al., "Measurementof the Differential Group Delay in Installed Optical Fibers Using Polarization Multiplexed Solitons," *IEEE Phot.Tech. Lett.*, **11**, 593-595 (1999).

Beals, K.J., Day CR, Duncan WJ, Dunn AG, Dunn PL, Newns GR and Wright JV, 5th European Conference on optical Fiber Communication (New York: McGraw Hill), 1979.

Boas, Mary L., *Mathematical Methods in the Physical Sciences* (Wiley, New York, 1983) p.454.

CCITT Recommendation G.651 - Characteristics of a 50/126 micron graded index optical fibre cable. *COM. XV-R-89 E,* May 1992.

Clark Jones, R., "A New Calculus for the Treatment of Optical Systems: VI. Experimental Determination of the Matrix," *J.O.S.A.* **37**, 110-112 (1946).

Corsi, F., A. Galtarosa and L. Palmieri, "Polarization Mode Dispersion Characterization of Single-Mode Optical Fibers Using Backscattering Technique", *Journal of Light wave technol.*, vol. 16, no. 10, pp. 1832 - 1843, 1998

Cristani, *Opt: Communn.*, Vol.146, 1998

Cyr, Normand *et al.*, "Stokes Parameter Analysis Method, the Consolidated Test Method for PMD Measurements," *Proceedings Nat. Fib. Opt. Eng. Conf.*, Chicago, (1999).

Derickson, D., Editor. Fiber Optic Test and Measurement, Upper Saddle River, New Jersey 07458, Prentice-Hall PTR 1998.

Fleming, J.W., "Fiber Optics Technicians Hand Book" by JIM Hayes, Delmar Publ. New York, Electron Lett. 14, (326), 1978.

Gisin, N. and Huttner, B. Combined effects of polarization mode dispersion and polarization dependent losses in optical fibers. *Opt. Commun.*, 1997, **142**, 119 - 125.

Gisin, N. and Pellaux, J.P. Polarization mode dispersion: time versus frequency domains. *Opt. Commun.*, 1992, **89**, 316 323.

Gisin, N., "Polarization Mode Dispersion: Definitions, Measurements and Statistics", OFMC' 93, Torino, pp. 149 - 154, 1993

Gisin, N., R. Passy, J. C. Boschoff and B. Perny, *IEEE Photon. Tech. Lett.*, vol. 5, p. 819, 1993.

Gisin, N.J., P. Von der Weid, J. P. Pellaux, "Polarisation mode dispersion of short and long single-mode fibers," Journal of Light wave Technology, Vol. 9, No. 7, pp. 82 1-827, JUIY 199 1.

Goodman, Joseph W., *Statistical Optics*,(Wiley, New York, 1985) p. 168.

Hecht, E. Optics, USA, Addison-Wesley, 1987 (2nd edition).

Hecht, Jeff, "The Story of Fiber Optics", Oxford University Press, New York, 1999.

Heffner, B.L, "Influence of optical source characteristics on the measurement of polarization-mode dispersion of highly mode-coupled fibers", *Opt. Lett.*, **21**, 113- 115, (1996).

Heffner, B.L., "Automated Measurement of Polarization Mode Dispersion Using Jones Matrix Eigenanalysis," *IEEE Photonics Tech. Lett.*, **4**, 1066-1069 (1992).

Heffner, B.L., "Compensation formula for noise threshold bias of interferometric PMD measurement," in NIST Technical Digest: Symposium on Optical Fiber Measurements 1996, Vol. 905, pp. 135-138, 1996.

Heffner, B.L., Hernday, R., Measurement of polarization-mode dispersion. *Journal of Hewlett-Packard.* 1995; 46(1): 27-33.

Heismann, F. (1998) *ECOC'98 Digest* **2** (Sept.), 51–79.

Huff RG, Dimarcello FV and Hast Jr AC, *Tech. Digest* OFC Paper TUG 2, 1989.

Huttner, B and GISIN, N. Anomalous pulse spreading in birefringent optical fibers with polarization-dependent losses. *Opt. Lett.*, 1997, **22**, 8, 504 – 506.

IEC 60793-1-48: to be published. Polarisation mode dispersion measurement methods.

IEC 61300-3-32: to be published. Polarisation mode dispersion for passive optical components.

Ihishi, V., Fujiura K. and Takahashi S., MRS Symp. Proc, 1990.

Izawa, T. and Sudo S., *Optical Fiber Materials and Fabrication* (36), 1987.

Izawa, T., Shabita N. and Takeda A., *Appl. Phys. Lett.* 3133, 1977.

Izawa, T., Sudo S., and Hanawa F ., Trans. IEEE 62, (779) 1979.

Jopson, R.M., "Measurement of Second- Order Polarization-Mode Dispersion Vectors in Optical Fibers," *IEEE Photonics Tech. Lett.* **11**, 1153-1155 (1999).

Lomair, P.J., Kranz, K.S., Walker, K.L. and Huff, R.G., *Elect. Lett.* 24(132), 1988.

Lu, K. E., Lee, M.T., Powers, D.R. and Glassman, S., in *Tech. Digest*, PDI,OFC (425), 1989.

MacChesney, J.B., Proc. IEEE 68(1181), 1980.

Menyuk, C. R. (1999) *J. Eng. Math.* **36**, 113–136.

Namihira, Yoshinori and Jun Maeda,"Polarization mode dispersion measurements in optical fibers", *Tech. Dig. – Symp.on Opt. Fiber Meas.*, Boulder, 145-150 (1992).

Nelson, Lynn E. *et al.* "Measurement of polarization mode dispersion vectors using the polarization-dependent signal delay method," *Opt. Express*, **6**, 158-167 (2000).

Obserson, P., Juliard, K., Gisin, N., Passy, R. and Weid, J., Interferometric polarization mode dispersion measurement with femtoseconds sensitivity. *Journal of Light wave Technology.* 1997; 15(10): 1852-1857.

Papamichel, H. and Miaoulis *J., MRS Symp. Proc.* 172(43), 1990.

Paul, France, Caster SF Moore W. and Williams JR in "Optical Fiber Material and Processing", MRS Symp. Proc. 172 (134), 1990.

Poole, C. D. and Nagel, J. A. (1997) in *Optical Fiber Telecommunications*, eds. Kaminow, I. P. & Koch, T. L. (Academic, San Diego), Vol. IIIA, pp. 114–161.

Poole, C. D. and Wagner, R. E. (1986) *Elect. Lett.* **22**, 1029–1030.

Poole, C.D. and C.R. Giles, "Polarization-dependent pulse compression and broadening due to polarization dispersion indispersion-shifted fiber", *Opt. Lett.*, **13**,155-157 (1987).

Poole, C.D. and Favin, D.L. Polarization-Mode Dispersion Measurements Based on Transmission Spectra Through a Polarizer. *J. Lightwave Tech.*, 1994, **12**, 6, 917 - 929.

Poole, C.D. and Wagner, R.E. Phenomenological approach to polarisation dispersion in long single-mode fibers. *Electron. Lett.*, 1986, **22**, 19, 1029 - 1030.

Poole, C.D. *et al.*, "Polarization dispersion and principal states in a 147 km undersea lightwave cable," *J. of Lightwave Tech.*, **LT-7**, 1185-1190 (1989).

Poole, C.D. *et al., Optics Letters*, vol. 13, pp. 155 - 157, 1988

Poole, Craig D., "Polarization-Mode Dispersion Measurements Based on Transmission Spectra Through a Polarizer," *J. Lightwave Tech.*, **12**, 917-929 (1994).

Prozean, K., *Proc. Indian Nat. Sci Acad.* 67, 2001.

Rashleigh, S. C. & Ulrich, M. (1978) *Opt. Lett.* **3**, 60–62.

Senior, John M., *Optical Fiber Communications,* Printice-Hall International (UK) Ltd. 1992.

TIA/EIA FOTP-122. 1999. *Polarization-Mode Dispersion Measurement for Single-ModeOptical Fibers by Stokes Parameter Evaluation.* Telecommunications Industry Association, Arlington, VA.

TIA/EIA FOTP-124. 1999. *Polarization- mode dispersion measurement for single- mode optical fibers by interferometric method.* Telecommunications Industry Association, Arlington, VA.

Turistyn, S.K. and E.G.Shapiro. *Opt. Lett* 23, 1998.

Williams, A. and C.M. Wang, "Corrections to Fixed Analyzer measurements of Polarization Mode Dispersion," *Journal of Lightwave Technology,* **16**, 534-541 (1998).

Williams, P.A. "Modulation phaseshift measurement of PMD using only four launched polarization states: a new algorithm," *Elec. Lett.*, **35**, 1578-1579 1999).

Williams, P.A. *et al.*, "Narrowband measurements of polarization-mode dispersion using the modulation phase shift technique," *Tech. Dig. - Symp. on Opt. Fib. Meas.*, Boulder, 23-26 (1998).

Index